Springer Theses

Recognizing Outstanding Ph.D. Research

Aims and Scope

The series "Springer Theses" brings together a selection of the very best Ph.D. theses from around the world and across the physical sciences. Nominated and endorsed by two recognized specialists, each published volume has been selected for its scientific excellence and the high impact of its contents for the pertinent field of research. For greater accessibility to non-specialists, the published versions include an extended introduction, as well as a foreword by the student's supervisor explaining the special relevance of the work for the field. As a whole, the series will provide a valuable resource both for newcomers to the research fields described, and for other scientists seeking detailed background information on special questions. Finally, it provides an accredited documentation of the valuable contributions made by today's younger generation of scientists.

Theses are accepted into the series by invited nomination only and must fulfill all of the following criteria

- They must be written in good English.
- The topic should fall within the confines of Chemistry, Physics, Earth Sciences, Engineering and related interdisciplinary fields such as Materials, Nanoscience, Chemical Engineering, Complex Systems and Biophysics.
- The work reported in the thesis must represent a significant scientific advance.
- If the thesis includes previously published material, permission to reproduce this must be gained from the respective copyright holder.
- They must have been examined and passed during the 12 months prior to nomination.
- Each thesis should include a foreword by the supervisor outlining the significance of its content.
- The theses should have a clearly defined structure including an introduction accessible to scientists not expert in that particular field.

More information about this series at http://www.springer.com/series/8790

Katherine A. Schreiber

Ground States of the Two-Dimensional Electron System at Half-Filling under Hydrostatic Pressure

Doctoral Thesis accepted by Purdue University, West Lafayette, Indiana, USA

 Springer

Katherine A. Schreiber
National High Magnetic Field
Laboratory - Pulsed Field Facility
Los Alamos National Laboratory
Los Alamos, NM, USA

ISSN 2190-5053   ISSN 2190-5061   (electronic)
Springer Theses
ISBN 978-3-030-26324-9   ISBN 978-3-030-26322-5   (eBook)
https://doi.org/10.1007/978-3-030-26322-5

This Springer imprint is published by the registered company Springer Nature Switzerland AG.
The registered company address is: Gewerbestrasse 11, 6330 Cham, Switzerland

Supervisor's Foreword

I am very pleased to introduce this work by my former student, Dr. Katherine A. Schreiber. In her thesis, Katherine describes cryogenic measurements at high hydrostatic pressures of the two-dimensional electron gas. Due to its simplicity and its richness, the two-dimensional electron gas remains one of the most interesting model systems in condensed matter physics. There are numerous electron gases known, but due to their exceptional quality, the ones confined to GaAs/AlGaAs heterostructures have historically produced several discoveries and continue to yield notable results.

Clean two-dimensional electron gases are best known for fractional quantum Hall states. These ground states are prototypical topological phases. In addition, the electron gases may host different charge-ordered Landau phases. The electron gas in GaAs/AlGaAs hosts is especially rich in such charge-ordered phases; examples are the Wigner solid and the electronic nematic and bubble phases.

In her work, Katherine explored the region of the phase space commonly referred to as the second Landau level, a region currently under intense scrutiny because of the numerous competing electronic phases, some of which are fractional quantum Hall states of unusual topological order. When tuned by hydrostatic pressures of the order of 10,000 atmospheres, the $\nu = 5/2$ and $\nu = 7/2$ fractional quantum Hall states give way to the nematic phase. The importance of these quantum phase transitions is twofold. First, these transitions are examples of rare phase transitions between a topologically ordered and a traditional Landau phases. Second, the fractional quantum Hall state involved is a very special one: it is due to the pairing of composite fermions, the emergent particles of the fractional quantum Hall regime, and it is thought to harbor non-Abelian excitations. Therefore, the phase transition found by Katherine highlights the competition of paired and nematic phases in the two-dimensional electron gas. The competition of pairing and nematicity is of current interest in several strongly correlated electron systems; observations of such a competition in the two-dimensional electron gas demonstrate that there is an intimate connection between the two phases involved.

This thesis provides an excellent introduction to quantum Hall physics, describes details of the high-pressure experiments, and discusses the ramifications and the origins of the newly reported phase transitions.

Professor of Physics   Gábor A. Csáthy
Department of Physics and Astronomy
Purdue University
West Lafayette, IN, USA
May 2019

Acknowledgments

The road to achieving my Ph.D. has been a long one, and I have many people to thank. First, I would especially like to thank my advisor, Dr. Gábor Csáthy. His investment in his graduate students' success is unmatched. Without the many hours he has spent giving me help and advice, I could not have achieved everything I have accomplished at Purdue.

I am indebted to my senior labmates whose advice was instrumental to my success in the lab: Nodar Samkharadze, Nianpei Deng, and Ethan Kleinbaum. In particular, I thank Nodar for initiating the pressure cell experiments and Ethan for his great patience and hard work helping me learn the ropes of low-temperature measurements. I also thank my junior labmates, Kevin Ro and Vidhi Shingla, for their help and support, often staying late in the lab to lend me a hand with my own experiments.

Keith Schmitter has been a huge help to me and the entire physics department, keeping our experiments afloat by providing us with liquid helium. Jim Corwin's advice and help in the machine shop have also been extremely helpful.

I thank several members of the Manfra group at Purdue. Dr. Mike Manfra, Saeed Fallahi, Dr. Geoff Gardner, Dr. John Watson, Dr. Tailung Wu, Qi Qian, Jimmy Nakamura, and Mike Yannell have all helped me with experimental problems at different points in my Ph.D. research, lending an extra hand, equipment, or advice. Dr. Manfra and Geoff Gardner have provided many of the high-quality samples central to my experiments. I also acknowledge Dr. Loren Pfeiffer and Dr. Ken West at Princeton University for samples.

I would like to thank my committee, Dr. Tongcang Li, Dr. Yuli Lyanda-Geller, and Dr. Mike Manfra, for the effort and time they have put into helping me achieve my degree. I would like to also acknowledge the theoretical help provided by the coauthors of my publications, Dr. Rudro Biswas, Dr. Eduardo Fradkin, and Dr. Yuli Lyanda-Geller.

Finally, I especially thank my sister, Rebecca, who has been a great source of inspiration and support. My boyfriend, Jeremy Prewitt, has been nothing but encouraging and patient throughout my many hours spent working toward this degree. Last of all, I have to thank my parents for supporting me and encouraging my interest in science throughout my life, taking me to science museums, letting me go to summer camps, and teaching me about Ohm's law and air conditioners. I could not have achieved my success without their help.

I acknowledge the US Department of Energy grant award DE-SC0006671 for funding this work. I would also like to acknowledge the Purdue University Cagiantas Fellowship for funding during the last year of my research.

Contents

1 The Quantum Hall Effect 1
   1.1   Two-Dimensional Electron Systems 1
   1.2   Classical Hall Effect 4
   1.3   Two-Dimensional Electron Systems in a Magnetic Field 6
   1.4   Integer Quantum Hall Effect 8
   1.5   Fractional Quantum Hall Effect 11
        1.5.1   Quasiparticles in the Fractional Quantum Hall Effect:
               Fractional Charge and Fractional Statistics 14
        1.5.2   The Composite Fermi Sea at $\nu = 1/2, 3/2$ 15
        1.5.3   The Quantum Hall Effect and Topological Order 15
   1.6 $\nu = 5/2$ Fractional Quantum Hall State 16
        1.6.1   Current Experimental Status of the $\nu = 5/2$
               Fractional Quantum Hall State 18
        1.6.2 $\nu = 7/2$ Fractional Quantum Hall State 21
   1.7   Conclusion 22
   References 22

2 The Quantum Hall Nematic Phase 25
   2.1   Nematicity in Condensed Matter Systems 25
   2.2   Prediction and Theory of the Nematic State in the
         Two-Dimensional Electron System 27
   2.3   Experimental Observation of the Nematic Phase:
         $\nu = 9/2, 11/2, 13/2$ 27
   2.4   The Effect of In-Plane Magnetic Field on the Nematic at
         $\nu = 9/2, 11/2, 13/2$ 28
   2.5   The Effect of In-Plane Magnetic Field on the Second Landau
         Level Fractional Quantum Hall States 29
        2.5.1   Nematic Fractional Quantum Hall States: $\nu = 7/3$
               and $\nu = 5/2$ 30
   2.6   Recent Studies of the Nematic Phase 31
   2.7   Other Anisotropic Signatures in Even Denominator States 32

2.8   Electron Solids: Wigner Crystal and Bubble Phases 32

2.9   Summary of States at Half-Filling 33

2.10   Conclusion 34

References 35

3 Low Temperature Measurement Techniques 37

3.1   Dilution Refrigeration 37

3.2   Low Noise Electronics 41

3.3   Conclusion 42

References 42

4 The Quantum Hall Effect and Hydrostatic Pressure 43

4.1   Gallium Arsenide Under Pressure 43

4.2   Previous Experiments of the Fractional Quantum Hall Effect
       Under Pressure 46

4.3   Pressure Clamp Cell 47

      4.3.1   Diamond Anvil Cells 49

4.4   Preparing for Pressurization and Cooldown 50

      4.4.1   Mounting the Sample to Pressure Cell Feedthrough 50

4.5   Monitoring the Effect of Pressure 53

      4.5.1   Room Temperature Pressure Monitoring 53

      4.5.2   Low Temperature Pressure Monitoring 55

4.6   Conclusion 58

References 58

**5 The Fractional Quantum Hall State-to-Nematic Phase
Transition Under Hydrostatic Pressure** 61

5.1   Observation of the Fractional Quantum Hall State-to-Nematic
       Transition at $\nu = 5/2$ 62

5.2   Spontaneous Rotational Symmetry Breaking 65

5.3   Topology, Pairing, and the Nematic Phase 67

5.4   Finite Temperature Studies at $\nu = 5/2$ 68

5.5   Quantum Phase Transition from Nematic Phase to Fermi
       Fluid-Like Phase 73

5.6   Conclusion 74

References 75

**6 Universality of the Fractional Quantum Hall State-to-Nematic
Phase Transition at Half-Filling in the Second Landau Level** 77

6.1   Observation of the FQHS-to-Nematic Phase Transition
       at $\nu = 7/2$ 77

6.2   Finite Temperature Studies at $\nu = 5/2$ and $\nu = 7/2$ 82

6.3   Conclusion 88

References 88

**7   Origin of the Fractional Quantum Hall State-to-Nematic Phase
     Transition in the Second Landau Level**...................................... 91
     7.1   Tuning the Electron–Electron Interactions with Landau Level
            Mixing... 91
     7.2   Tuning the Electron–Electron Interactions Through Quantum
            Well Width 92
     7.3   The Role of Electron–Electron Interactions in the Fractional
            Quantum Hall State-to-Nematic Phase Transition 93
     7.4   Observation of the Nematic Phase at $\nu = 7/2$ at Ambient
            Pressure 96
     7.5   Recent Theory of the Transitions to the Nematic Phase 98
     7.6   Importance of the Second Landau Level for the
            FQHS-to-Nematic Phase Transition 98
     7.7   Conclusion 100
     References 100

Parts of this thesis have been published in the following journal articles

- N. Samkharadze, K.A. Schreiber, G.C. Gardner, M.J. Manfra, E. Fradkin, G.A. Csáthy, Nat. Phys. **12**, 191 (2016)
- K.A. Schreiber, N. Samkharadze, G.C. Gardner, R.R. Biswas, M.J. Manfra, G.A. Csáthy, Phys. Rev. B **96**, 041107 (2017)
- K.A. Schreiber, N. Samkharadze, G.C. Gardner, Y. Lyanda-Geller, M.J. Manfra, L.N. Pfeiffer, K.W. West, G.A. Csáthy, Nat. Commun. **9**, 2400 (2018)

Chapter 1
The Quantum Hall Effect

Electrons in two-dimensional semiconductor structures subjected to a magnetic field have yielded a wealth of diverse electronic ground states. Perhaps the most famous class of these ground states are the integer quantum Hall states (IQHSs) and fractional quantum Hall states (FQHSs). In this chapter, I introduce the quantum Hall effect and describe a state of particular interest, the $\nu = 5/2$ fractional quantum Hall state. For a detailed introduction to the two-dimensional electron system and the quantum Hall effect, I refer the reader to the textbooks and reviews in references [1–5].

1.1 Two-Dimensional Electron Systems

Reduced dimensionality has permitted the observation of novel quantum effects, leading to some of the most exciting recent discoveries in condensed matter physics. The development of high mobility heterostructures hosting two-dimensional electron systems (2DESs) paved the way for many of these discoveries. Not only does a 2D system display many quantum phenomena in its own right, lithography techniques on 2D systems permit the relatively facile creation of 1D systems (nanowires) and 0D systems (quantum dots). For these reasons, the 2DES continues to be a fundamental system for hosting new physics.

The 2DES was among the first low dimensional systems to be realized. The accumulation region of a silicon MOSFET, for example, was an early manifestation of the 2DES, and boasts the first observation of the integer quantum Hall effect [1, 6]. GaAs heterostructures proved to be a very high mobility system that permitted the observation of even more fragile states, such as the fractional quantum Hall effect [7]. Graphene, the celebrated carbon 2D material, is particularly exciting because of its Dirac dispersion, leading to massless Dirac fermions [8–10]. Fabrication techniques in graphene are ever improving, and have now led to the observation

© Springer Nature Switzerland AG 2019

K. A. Schreiber, *Ground States of the Two-Dimensional Electron System at Half-Filling under Hydrostatic Pressure*, Springer Theses,
https://doi.org/10.1007/978-3-030-26322-5_1

of fractional quantum Hall states in this material [11]. Other notable 2DESs can be found on the surface of superfluid helium [12], in ZnO heterostructures, [13], and in thin layered Van der Waals materials, such as the transition metal dichalcogenides [14].

The GaAs system remains an extremely high quality system for observing quantum effects. Indeed, the discoveries of most of the interesting 2D electron states belong to GaAs. Several features of GaAs heterostructures contribute to their excellent propensity for revealing novel electron states. These heterostructures are made from junctions of GaAs with $Al_xGa_{1-x}As$, where x is the concentration of Al. The concentration of Al, x, may be tuned in growth to achieve the appropriate barrier height for a confining potential of the 2DES, and a typical value is some $x = 20$–30% [1, 15, 16]. GaAs and $Al_xGa_{1-x}As$ have very similar lattice constants, permitting relatively smooth interfaces to form, reducing interface scattering in the 2DES [1]. Such interfaces are readily grown by molecular beam epitaxy (MBE) [15, 16]. In an MBE chamber, beams of atoms from a heated reservoir of the desired element are deposited layer by layer onto a substrate. MBE is the standard technique for growing the samples we have measured. GaAs heterostructures as such have a high mobility. Mobility is given by $\mu = e\tau/m$, where τ is the scattering lifetime. A sample with a high mobility has a long scattering lifetime and therefore a large mean free path. Contemporary samples can have a mobility on the order of $10^7 \, cm^2/Vs$, which corresponds to a mean free path of several hundred microns [15, 16].

Single heterojunction samples of $GaAs/Al_xGa_{1-x}As$ were the early standard heterostructure types for studying 2DES physics, with the 2DES just at the interface of these two materials. An important development in improving the sample quality of these heterostructures was the innovation of modulation doping [16]. Modulation doping involves the placement of dopant atoms remotely from the 2DES region. For n-type samples, the dopant is silicon. The dopant is placed in a narrow well a few nanometers wide, some 50–100 nm away from the 2DES. This reduces the effect of the ionized dopant atoms on scattering electrons in the 2DES.

The development and refinement of the GaAs quantum well structure improved 2DES quality further. The quantum well consists of a layer of GaAs, usually between 20 and 60 nm, sandwiched between $Al_xGa_{1-x}As$ layers. This symmetric structure permits modulation doping from both sides of the quantum well, increasing the carrier density in the quantum well while maintaining the dopant setback at the desired distance. Additionally, scattering from the $GaAs/Al_xGa_{1-x}As$ interfaces is reduced, as when the lowest quantum well energy level is occupied, the 2DES forms well-centered between the well walls. The conduction band minimum profile of a typical quantum well is given in Fig. 1.1a [15], and a close-up of the doping well structures are seen in Fig. 1.1b. A picture of such a GaAs sample mounted in a measurement header is depicted in Fig. 1.1c.

Finally, 2DESs in $GaAs/Al_xGa_{1-x}As$ have benefited from improvements in molecular beam epitaxy technique [15, 16]. Ultrahigh vacuum is mandatory for reducing impurity levels in the sample sufficiently. Cryopumps are therefore needed for such a vacuum, and extensive baking of MBE components is needed to bake off impurities. Careful choice of MBE components is needed generally to

Fig. 1.1 (a) The conduction band edge minimum (red) of the GaAs/Al$_x$Ga$_{1-x}$As quantum well heterostructure, and the electron density profile (blue). The electrons are concentrated at the peak within the quantum well, near the position 210 nm. Reprinted with permission from Ref. [15]. (b) A close-up of a delta doping well, doped with silicon. Reprinted with permission from Ref. [15]. (c) A photograph of a 2 × 2 mm GaAs sample mounted on a header for measurement

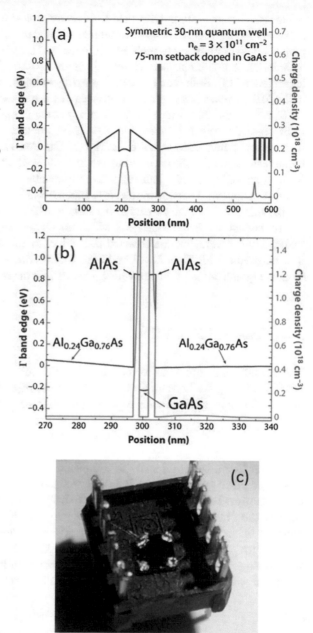

reduce outgassing—eliminating the use of many typical plastic and polymer sealing materials as well as lubricants. Reference [16] demonstrates the first sample in which an extremely high mobility of over $10^7 \, \text{cm}^2/\text{Vs}$ was attained by MBE techniques. A comprehensive review of GaAs/$\text{Al}_x\text{Ga}_{1-x}$As MBE growth is given in reference [15]. Both works emphasize stringent attention to detail and cleanliness in the MBE plays a large role in the growth of the highest quality samples.

To access the 2DES in a GaAs sample, ohmic contacts are needed. Typically, In/Sn eutectic solder is used, but Au/Ge/Ni contacts may be used, especially for contacts patterned by photolithography. The contacts are annealed in a small homemade annealing furnace in our lab at around $450 \, °C$ for a few minutes, in a forming gas of H_2 and N_2. This recipe usually results in good quality ohmic contacts that do not appear to impede the electron states forming in the 2DES. These contacts are the shiny blobs in the corners of the sample in Fig. 1.1c.

For the rest of this introduction, I will focus on the electronic phases of the GaAs 2DES. I will first review the classical Hall effect to detail the behavior of electrons in a magnetic field. Then I will build the formulation of the quantum Hall effect from the quantization of electron energy levels in a magnetic field.

1.2 Classical Hall Effect

The classical Hall effect has its roots in a familiar concept from classical electrodynamics [17, 18]: the Lorentz force on a moving charge. An electron moving with a velocity $\vec{v}$ in a magnetic field $\vec{B}$ experiences the force

$$\vec{F}_L = e\vec{v} \times \vec{B} \tag{1.1}$$

If we consider a current density $\vec{J} = ne\vec{v} = J\hat{y}$ in a material in the presence of a perpendicular magnetic field $\vec{B} = B\hat{z}$, the Lorentz force entails a separation of charge carriers in the direction transverse to both the current and the magnetic field (that is, the direction $\hat{x}$). Once the charges separate, if the magnetic field remains constant in time, the system reaches a steady state, and a charge carrier making up the current must then feel no net force in the $\hat{x}$ direction. There is an electric field arising from the separated charges $\vec{E}$ that exactly balances the Lorentz force: $q\vec{E} = q\vec{v} \times \vec{B}$. Using the coordinate system we have set up, this means

$$E_x = v_y B_z = v_y B \tag{1.2}$$

This transverse electric field gives rise to a voltage drop across the sample, known as the Hall voltage, from which we can extract the Hall resistivity. The resistivity tensor makes its appearance in the precursor of Ohm's law:

$$\vec{E} = \bar{\rho}\vec{J} \tag{1.3}$$

In the absence of a magnetic field, $\bar{\rho}$ is diagonal: $E_x = \rho_{xx} J_x$ and $E_y = \rho_{yy} J_y$. ρ_{xx} and ρ_{yy} are referred to as the longitudinal resistivity, and are about equal for an isotropic material. Introducing a perpendicular magnetic field as above, however, adds off-diagonal elements to the resistivity tensor. Since the definition of current density is $\vec{J} = ne\vec{v}$, where n is the density of electrons, we can see how the Lorentz force comes into play. Plugging in for the transverse electric field, we see $E_x = B J_y/ne$. Following the rules of the cross product, we also see $E_y = -B J_x/ne$. The resistivity tensor then gives

$$\bar{\rho} = \begin{pmatrix} \rho_{xx} & B/en \\ -B/en & \rho_{yy} \end{pmatrix} \tag{1.4}$$

The component $\rho_{xy} = B/en$ is the *Hall resistivity*, and does not depend on any properties of the material except for carrier density and the sign of the carriers. Hence, Hall effect measurements are the standard way of determining the carrier density in new materials, and whether the carriers are electrons or holes [18]. We can extend the discussion from resistivity to resistance by multiplying by the appropriate geometrical factors. The measurement of Hall resistance R_{xy} is easily done in a four-terminal contact setup, as pictured in Fig. 1.2, sourcing the current through the sample, applying the magnetic field, and measuring the voltage drop transverse to the current. The longitudinal resistance R_{xx}, the resistance along the direction of

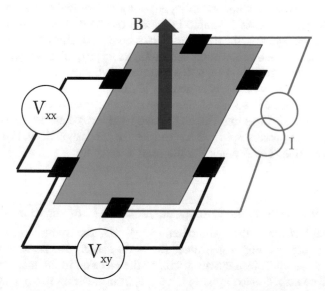

Fig. 1.2 The basic setup of a transport measurement for obtaining Hall resistance and longitudinal resistance. A current is passed through the sample, and a magnetic field B is applied perpendicular to the current. By measuring the voltage drop along the direction of current and dividing by the current, we obtain the longitudinal resistance, $R_{xx} = V_{xx}/I$. By measuring the voltage drop transverse to the current and dividing by the current, we obtain the Hall resistance $R_{xy} = V_{xy}/I$

current, is also measured easily in a four-terminal setup. In the presence of magnetic field, it is often referred to as magnetoresistance.

Importantly, the classical Hall resistance is strictly linear in the magnetic field, giving a Hall slope of $1/en$, and the classical magnetoresistance is finite, reflecting the scattering lifetime of carriers in the material. We shall see that this behavior does not hold in the case of the quantum Hall effect.

1.3   Two-Dimensional Electron Systems in a Magnetic Field

The quantum Hall effect epitomizes the reflection of fundamentally quantum behavior in electrical transport. To understand the quantum Hall effect, we first need to understand the quantum mechanical behavior of electrons in a magnetic field. The Hamiltonian for an electron in a uniform magnetic field $\vec{B}$ with a vector potential $\vec{A}$ and some generic scalar potential V is

$$\mathcal{H} = \frac{(\vec{p} - e\vec{A})^2}{2m} + V \tag{1.5}$$

[19]. Taking the direction of the magnetic field $\vec{B}$ to lie on the z-axis, so that $\vec{B} = B\hat{z}$, we choose the vector potential to be $\vec{A} = Bx\hat{y}$. This choice of vector potential yields the energy spectrum in an elegant way, and is referred to as the Landau gauge. When we choose this gauge, the Hamiltonian can be written in a form identical to that of a simple harmonic oscillator Hamiltonian [1, 2, 19]. The Schrödinger equation for the simple harmonic oscillator can be solved using the concept of raising and lowering operators, as in [19]. The energy eigenvalues are the important result: we obtain a spectrum of equally spaced Landau levels: $E_j = \hbar\omega_c(j + 1/2)$, where j is an integer [1, 19]. ω_c is the cyclotron frequency: $\omega_c = eB/m$. It should be noted here that for electrons moving within a solid material, m is the effective mass. Each Landau level has a degeneracy $D = eB/h$.

Electrons also possess spin, which couples to the magnetic field through the Zeeman interaction. The Zeeman Hamiltonian is given by

$$\mathcal{H}_Z = -\vec{\mu} \cdot \vec{B} \tag{1.6}$$

where $\vec{\mu}$ is the magnetic dipole moment, $\vec{\mu} = \frac{\mu_B g \vec{S}}{\hbar}$. In our case, in which the magnetic field is along the z-direction, $\vec{B} = B\hat{z}$, the Hamiltonian is $\mathcal{H}_Z = -\frac{\mu_B g B}{\hbar} S_z$, giving us energy eigenvalues $E_\pm = \pm\frac{1}{2}g\mu_B B$. The Zeeman energy also contributes to the quantized energy levels of the electron in the magnetic field, so that the total contribution to the energy due to spinful electrons in a magnetic field is. Here I am simply denoting μ_B as μ for clarity.

$$E_{j,\pm} = \hbar\omega_c\left(j + \frac{1}{2}\right) \pm \frac{1}{2}g\mu B \tag{1.7}$$

This Zeeman splitting is observable in the quantum Hall effect in GaAs, as discussed below, with a Landé g-factor of $g = -0.44$. It should be noted that in materials such as graphene, valley degeneracy lifting further plays a role in splitting the Landau levels, leading to a fourfold splitting in graphene's case [8–11]. I will focus only on GaAs, but generally the energy levels may reflect the lifting of other various degeneracies besides spin.

In a two-dimensional electron system (2DES), these quantized energy levels are observable in transport, while in a three-dimensional material, they are not. To see why, consider first an electron free to move in all three dimensions, experiencing the uniform magnetic field $\vec{B} = B\hat{z}$. The electron is a free particle in the z-direction while completing cyclotron orbits in the $x - y$ plane. Its energies are therefore

$$E_{j,\pm,k_z} = \hbar\omega_c \left(j + \frac{1}{2} \right) \pm \frac{1}{2}g\mu B + \frac{\hbar^2 k_z^2}{2m} \tag{1.8}$$

However, k_z is a continuous variable, so this means the energies of the electron in three dimensions are continuous. Indeed, in transport measurements, we do not observe signs of quantization.

In a 2D system, the electrons are to a good approximation confined to move in a single plane, the $x - y$ plane. The sample heterostructure confines the electrons in the z-direction. The most recent high quality GaAs heterostructures are quantum wells about 30–60 nm wide. We are able to approximate the energy of the electrons in this quantum well by the energies of the infinite square well, $E_{\alpha, ISQ} = \frac{\alpha^2 \pi^2 \hbar^2}{2mw}$, where α is an integer, and w is the width of the well. In heterostructures, this is called the *first subband*. When we do not make the infinite square well approximation, and use the true potential profile of the quantum well, or even use a triangular well approximation as in the case of single heterojunction samples, quantized energy levels still arise. We still refer to the levels of this confining potential as subbands. I will label the energy of the subbands for a general potential by E_α.

For low enough densities and narrow enough quantum wells, only the lowest square well level is occupied. In practice, this is usually desirable. Samples of densities low enough that only the first subband is occupied (that is, so that the Fermi level lies between the first and second subbands) are of highest mobility. This is because the first subband wavefunction, even in the real case where we do not make the infinite square well approximation, has a single local maximum, confining most of the electrons to the center of the quantum well. The second subband has a node at the well center and two maxima near the well edge—meaning most electrons are closer to the edge, and will scatter off of the rough interface more.

We can see now that the energy in a 2DES is truly quantized. Assuming that we are always populating only the first quantum well subband, $\alpha = 1$, as is the case for the samples studied in this thesis, the electrons have the energy

$$E_{j,\pm,\alpha=1} = \hbar\omega_c (j + \frac{1}{2}) \pm \frac{1}{2}g\mu B + E_1 \tag{1.9}$$

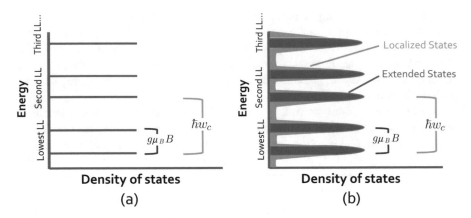

Fig. 1.3 (**a**) The spin-split Landau levels. The Landau levels are separated by the cyclotron energy, $\hbar\omega_c$, and each Landau level is split by the Zeeman energy splitting, $g\mu_B B$. There are therefore two spin branches per Landau level. (**b**) Disorder broadens these energy levels, giving rise to extended states which contribute to conduction (blue), and localized states which do not (orange)

One more ingredient is needed to make this description more realistic: disorder. In Fig. 1.3a, the spin-split Landau levels in the lowest subband are depicted exactly, as delta functions with a large density of states reflecting the levels' degeneracy of $D = eB/h$. In reality, disorder broadens these energy levels, as depicted in Fig. 1.3b. Two types of states arise from the inclusion of disorder. The first type of states are the extended states, depicted in dark blue. These represent energies slightly different from the spin-split Landau level energy due to the influence of defects, broadening the level. These defects do not inhibit the ability of the electrons to conduct through the sample. The orange states in the figure are called the localized states. These reflect the energies of electrons that are trapped by defects, and do not contribute to conduction. The disorder is characterized by a characteristic energy $\Gamma = \hbar/\tau_i$, where τ_i is the quantum lifetime, the average time between an electron's scattering events. To resolve the Landau levels, the condition $\hbar\omega_c > \Gamma$ is necessary.

As we shall see, these disorder-broadened, quantized energy levels become reflected in the transport properties of the 2DES. The quantum Hall effect is the manifestation of the changing population of these levels with changing magnetic field.

1.4 Integer Quantum Hall Effect

The first class of electronic states unique to the two-dimensional electron system are the integer quantum Hall states (IQHSs). The integer quantum Hall effect (IQHE) was discovered by von Klitzing in 1980, as he applied a magnetic field to a silicon MOSFET [6]. As the magnetic field was increased, the linear increase of the classical Hall resistance was observed, accompanied by very flat plateaus perfectly quantized, where $R_{xy} = \dfrac{B}{ne} = \dfrac{h}{je^2}$ for j an integer. This integer j was realized to relate to a quantity $v = \dfrac{h}{e}\dfrac{n}{B}$. The plateaus in R_{xy} extended over a short range of magnetic field, encompassing the magnetic field at which v is an integer. v is known as the *filling factor*, and as we shall see, it represents how many available electron states of a spin-split Landau level are filled. At the same magnetic field values at which the Hall plateaus arise, there are also dramatic minima in the longitudinal resistance. These can be seen in the top plots of Fig. 1.4.

To understand the origin of these plateaus and minima, it is productive to think of the extended states of the spin-split Landau levels as bands separated by gaps. The odd integer bands—the lower spin branches of each Landau level—have a gap equal to the Zeeman energy, and the even integer bands—the upper spin branches—have a gap of the cyclotron energy. The bottom plots of Fig. 1.4 make use of this analogy. The extended states are denoted by dark blue for filled levels and light blue for unfilled levels, and the localized states are colored dark orange for filled and light orange for unfilled.

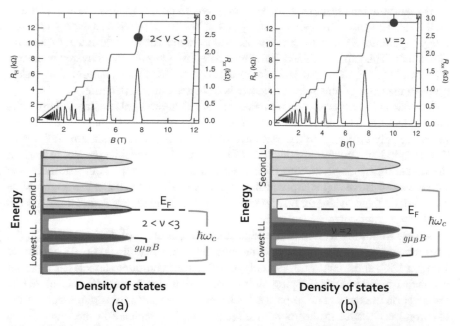

Fig. 1.4 The spin-split Landau levels in a GaAs 2DES in a magnetic field. (**a**) Imagining the spin-split Landau levels as bands, one has normal conduction when the Fermi level lies within an energy level (blue states). (**b**) Increasing the magnetic field, we observe quantized conduction as long as the Fermi level lies in a gap, where the electrons are localized and do not contribute to conduction (orange states). IQHE plot adapted from [20]

When the Fermi energy lies inside the extended states of a spin branch of a Landau level, there are electron states that may contribute to conduction. In this case, the resistances we observe are the classical linear Hall resistance and a finite longitudinal resistance, as we would observe for a three-dimensional sample (Fig. 1.4a). When the magnetic field increases, the spacing of the levels increases and their degeneracy increases to accommodate more electron states. Eventually the Fermi level lies in a gap, where there are only localized states, and there are no states available for conduction. As long as the Fermi energy lies in the gap, the bulk will be insulating and the resistance will not change. This is an IQHS. We can discern from this that disorder is the reason that a quantum Hall state is observed over a range of magnetic field, rather than only briefly as the Fermi energy moves from one Landau level to the next. The more of the localized states there are, the greater the magnetic field range over which the Fermi energy lies in the gap, and the wider the plateaus and minima. Thus a perfectly pristine sample is *not* ideal for observing the quantum Hall states, because the width of the plateaus and minima will be too small to observe.

We now recall that each band has a degeneracy of eB/h—total magnetic field divided by number of flux quanta. We can now obtain a physical meaning for the filling factor $\nu = n/\frac{eB}{h}$ as well: it is the ratio of electron density to the density of available states, that is, the number of filled spin-split branches of Landau levels.

Because the bulk is gapped, the system is said to be incompressible at a quantum Hall state. The gap is, as depicted in Fig. 1.4, given by the cyclotron energy for even integer quantum Hall states and the Zeeman energy for the odd integer quantum Hall states. The gap is obtained from the Arrhenius equation for activated behavior, $R_{xx} \propto e^{-\Delta/2k_B T}$. The gap is an important characteristic of quantum Hall states, giving a measure of how robust the state is to increasing temperature.

The question arises: why do we measure a zero longitudinal resistance if no states are conducting? The above analysis pertains to the sample *bulk*, and the resistances measured rely critically on the existence of edge states in the quantum Hall states. The sample has a confining potential at its edges, which must be present to prevent the tunneling of electrons into the air surrounding the sample. As a result, each Landau band curves up sharply at the edges as in Fig. 1.5a. Thus there are points at the edge where the Fermi energy must cross the Landau levels, and these are the conducting edge states.

A very important feature of the edge states is that they are dissipationless in the zero temperature limit, thanks to time reversal symmetry breaking. As shown by Büttiker [21], at the edge of the sample, the electrons complete semiclassical partial orbits that result in their skipping motion forward along the edge (Fig. 1.5b). When the electron encounters an impurity, backscattering is strongly suppressed, but the electron may scatter in the forward direction also as depicted in the figure. As a result, the resistance will be very low along an edge state. We thus will measure nearly zero voltage drop between two contacts along the same edge state in a four-terminal measurement configuration.

Why do we measure the Hall resistance quantized at the values that they are? One may compute the conductance of an edge state from the definition of current

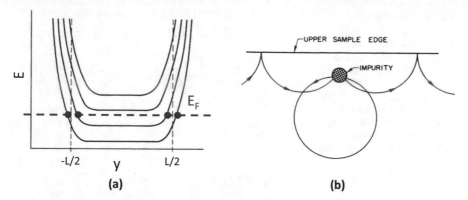

Fig. 1.5 (**a**) Edge states in a sample of length L. The Landau level energy bands curve at the sample edge due to the confining potential of the sample. The Fermi energy crosses these upturned bands at the edge, resulting in conducting edge states. Adapted from [21]. (**b**) Suppressed backscattering in an edge state. In a semiclassical picture, the electron completes skipping orbits along the edge. Impurities cause scattering into the forward direction only. This means we measure a minimum in longitudinal resistance at a quantum Hall state. Image from Ref. [21]. Reprinted figure with permission from M. Büttiker, Phys. Rev. B, 38, 9375 (1988). Copyright 1988 by the American Physical Society

and the density of states in one dimension. This may be done through Landauer–Büttiker formalism, and is described in Ref. [1] as well as by Büttiker in Ref. [21]. The conductance of each edge state is e^2/h, so at $\nu = 3$, for example, the edge states have total conductance $3e^2/h$. Therefore, in a four-terminal measurement, the voltage difference across a sample, transverse to the applied current, will be given by $V_{xy} = I \frac{h}{Ne^2}$, where N is the number of edge states. In this manner, we obtain the Hall resistance of $R_{xy} = h/Ne^2$.

1.5   Fractional Quantum Hall Effect

While the IQHE can be neatly explained by the energy spectrum of a single charged particle in a magnetic field, the fact that electrons interact with each other leads to more exciting and intricate behavior. In 1982, Tsui, Stormer, and Gossard astonishingly discovered a quantum Hall plateau and minimum at $\nu = 1/3$, which could not be explained in the Landau level picture above [7]. At $\nu = 1/3$ in the single particle Landau level picture, the Fermi level lies fully inside the lowest Landau level, which corresponds to an ungapped state. Soon afterward, it was clear that quantized states existed at *many* fractional values of ν. As it turns out, the Coulomb interaction between the electrons opens further gaps in the energy spectrum, and plays a major role in determining their behavior.

Fig. 1.6 The first observation of the $\nu = 1/3$ FQHS. A minimum in longitudinal resistivity ρ_{xx} and a quantized plateau in ρ_{xy} are seen near $B = 150\,\mathrm{kG}$ (bottom axis), at $\nu = 1/3$ (top axis). Ref. [7]. Reprinted figure with permission from D. C. Tsui, H. L. Stormer, and A. C. Gossard, Phys. Rev. Lett., 48, 1559 (1982). Copyright 1982 by the American Physical Society

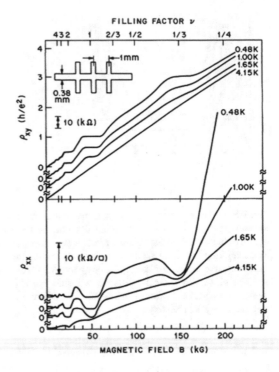

This problem is an example of a very complicated many-body problem: solving the Hamiltonian eigenvalue equation for $H_{Coulomb} = \sum_{i \neq j} \frac{e^2}{4\pi\epsilon|r_i - r_j|}$. The number of electrons summed over in this Hamiltonian is on the order of 10^{11}. Fortunately, using variational techniques, Laughlin [22] was able to write down a wavefunction that describe states at filling factors $\nu = 1/m$, where m is an *odd* integer.

$$\Psi_{1/m}^{\text{Laughlin}} = \prod_{j<k}(r_j - r_k)^m \exp\left[-\frac{1}{4l_B}\sum_i |r_i|^2\right] \tag{1.10}$$

Here, $l_B = \sqrt{h/eB}$ is the magnetic length, which accounts for the average distance between electrons at a given magnetic field. One can see that m must be odd, because the wavefunction must be antisymmetric under exchange of the particle's position (Fig. 1.6).

This wavefunction was generalized to account for numerators greater than one, building up what is called the hierarchy of states [23]. Indeed, the hierarchy of states described well the FQHSs observed: $\nu = 1/3, 2/5, 3/7, 4/9 \ldots$.

There is another intuitive way, put forward by Jain [3, 24], to understand the appearance of the fractional quantum Hall effect at the observed odd denominators. One considers the existence of particles called composite fermions (CFs), composed of an electron and an even number of quantized vortices. A vortex is here a spatially localized, quantized amount of magnetic field with half-integer spin, such that an

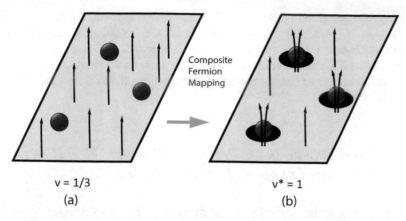

v = 1/3   v* = 1
(a)   (b)

Fig. 1.7 (a) A representation of electron filling factor $\nu = 1/3$, where the flux quanta are black arrows and the electrons are red circles. In this state, the electrons are strongly interacting. There is one electron for every three flux quanta. (b) *Composite fermion* effective filling factor $\nu^* = 1$, where the CFs are red circles with black arrows, and the effective magnetic field flux quanta are depicted by black arrows. There is one composite fermion for every one flux quantum, hence the effective filling factor is $\nu^* = 1$. The composite fermions are weakly interacting, so the FQHSs can be described by an effective integer quantum Hall effect of composite fermions

electron that completes a closed loop around it acquires a phase of 2π. The areal density of vortices at a given magnetic field is given by the areal density of flux quanta, $\frac{B}{h/e}$. We can deduce the behavior of a composite fermion by considering it as an electron with an even number of flux quanta "attached" to it. The flux quanta are depicted in Fig. 1.7 by arrows.

Now consider mapping a system of electrons to composite fermions as in the figure. Jain's realization was that we may treat the system of composite fermions as *non-interacting* particles in an *effective* magnetic field given by the remaining field after the flux attachment procedure is complete. For example, at filling factor $\nu = \dfrac{n}{eB/h} = 1/3$ there is one electron for every three flux quanta, shown in Fig. 1.7a. Performing the mapping by assigning two flux quanta to each electron, we have some flux quanta "left over." In fact we see we have one composite fermion for every one flux quantum (Fig. 1.7b). So we find that the effective filling factor $\nu^* = \dfrac{n}{eB^*/h} = 1$. In this effective magnetic field, the weakly interacting composite fermions are in an effective *integer* quantum Hall state, and therefore we see the quantized Hall plateau and longitudinal minimum. The energy levels of CFs are analogous in structure to the Landau levels, and are sometimes called lambda levels [3].

The composite fermion mapping works well for most observed FQHSs of odd denominator. The general relationship between filling factor and effective filling factor is given by $\nu = \frac{\nu^*}{2p\nu^*\pm 1}$, where $p = 1$ describes FQHSs obtained by mapping two quantized vortices to each electron, $p = 2$ describes FQHSs obtained by mapping *four* quantized vortices to each electron, and so forth. Furthermore, states such as $\nu = 2/3, 3/5$, and others greater than 1/2 are, for the most part, understood as the particle-hole conjugates of their cousins $\nu = 1/3, 2/5$, etc., and should be understood with the same composite fermion physics. As we shall see, however, the picture is not always so simple, and there exist fractions which do not follow the CF picture so neatly.

1.5.1   Quasiparticles in the Fractional Quantum Hall Effect: Fractional Charge and Fractional Statistics

When the filling factor is exactly one of the special fractions discussed above, and when the temperature is zero, the 2DES hosts exactly the FQHS ground state. Finite temperature and small magnetic field deviations away from an exact fractional filling factor result in the generation of quasiparticle excitations of the FQHS [2–4]. These quasiparticle excitations are exciting because they have very unusual properties: fractional charge and fractional statistics [5, 25–27].

Quasiparticles of a FQHS at $\nu = \frac{\nu^*}{2p\nu^*\pm1}$ have a charge of $q = \frac{e}{2p\nu^*\pm1}$ [22]. They are obviously not the result of an electron being literally divided, but rather they are complex effects of an interacting many-body electron system. A heuristic explanation for why the quasiparticles have fractional charge is given in reference [4], which I summarize here, viewing quasiparticles as defects in the CF ground state. If the system is tuned exactly to the ground state at $\nu = 1/m$, and another electron is added, defects are generated in the CF sea. To remain at the same filling factor, one would need m extra vortices: an even number $m - 1$ to combine with the electron, and one free, so that the ratio of CFs to vortices remains at $\nu^* = 1$. The dearth of these needed vortices is reflected in the defects created in the CF sea: exactly m quasiparticles with a total charge that must match the charge of the added electron. In this way, the quasiparticles are concluded to have a charge of $q = e/m$ at filling factors with denominator m. These fractional charges are in fact experimentally observable in shot noise experiments, discussed in some more detail later [28, 29].

These quasiparticles also have the unusual property of fractional statistics. A quasiparticle with fractional statistics behaves like neither a fermion nor a boson in 2D [5]. When one interchanges the position of two bosons, the wavefunction remains the same; when one interchanges the positions of a fermion, the wavefunction acquires a negative sign. When two quasiparticle excitations of fractional statistics—called anyons—are exchanged, the wavefunction acquires a factor of $e^{i\theta}$, where θ is not necessarily an integer multiple of π. Because the anyons are confined in 2D, their paths enclose magnetic flux when they are braided around each other in an exchange. This causes them to acquire a phase, which is reflected in their anyonic statistics [5, 25–27].

The energy to generate a quasiparticle excitation is precisely the gap of the FQHS, that is, the energy spacing between the CF lambda levels [3]. Like the IQHSs, the FQHSs are incompressible, and the gap is measurable by the Arrhenius equation.

1.5.2 The Composite Fermi Sea at $v = 1/2, 3/2$

The absence of a FQHS at $v = 1/2$ and $3/2$ is also explained by composite fermion theory. Each electron gets exactly two vortices, and none are left over, so the CFs behave as though they are in zero field. Halperin et al. [30], independently from Jain, indeed confirmed that a Fermi sea was theoretically expected at $v = 1/2$ and $v = 3/2$. Experimental signatures have been found for this Fermi sea as well. Kang et al. found signatures of cyclotron orbits of composite fermions at $v = 1/2$, much like what is observed at $B = 0$, which is a true Fermi sea of electrons [31]. Similarly, Du et al. [32, 33] found evidence for an unconventional Fermi sea at $v = 1/2$ made of composite fermions. Additionally, Willett et al. [34] found evidence for a Fermi surface at $v = 1/2$ using acoustic techniques. The resistance signature of the composite Fermi sea at $v = 1/2$ is a featureless trace (Fig. 1.8).

From the elementary composite fermion formalism of the FQHSs, it is logical that there is a composite Fermi sea at $v = 1/2$ and $v = 3/2$. However, in the higher Landau levels, there are striking contradictions to this rule. At $v = 5/2$ and $v = 7/2$, there are in fact fractional quantum Hall states, which I will describe in the following section. At $v = 9/2, 11/2, 13/2$, and so on, there is a broken symmetry phase called the nematic phase, to which I will devote the next chapter. Therefore, we have our first inkling that the states at half-filled Landau level spin branches are states of very special physics.

1.5.3 The Quantum Hall Effect and Topological Order

The IQHSs and the FQHSs fall into a category of phases known as topological phases [35, 36]. Topological phases have become a central theme in contemporary condensed matter research. A topological phase is one that consists of an insulating bulk with conducting edge states (for 2D materials) or surface states (for 3D materials). These conducting edge states arise due to a topologically protected configuration of the band structure. As the number of loops in a knot cannot be changed unless relatively significant energy is put into untying the knot, so the number of times the Fermi level crosses the band at the sample edge cannot be changed unless a great deal of energy is put into significantly changing the energy level structure. This robustness of the edge states has piqued the interest of researchers, not least because a protected edge state could be used in fault-tolerant quantum computing operations.

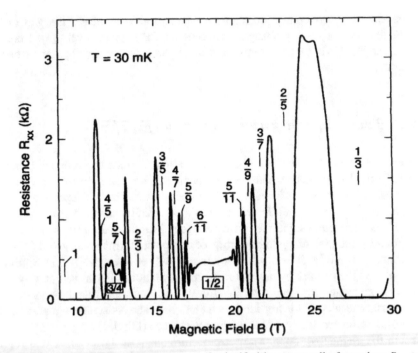

Fig. 1.8 The composite Fermi sea at $\nu = 1/2$, signified by a generally featureless R_{xx} trace at this filling factor. Compare with the sharp fractional quantum Hall state minima nearby. Ref. [33]. Reprinted figure with permission from R.R. Du et al. [33]. Copyright 1994 by the American Physical Society

1.6 $\nu = 5/2$ Fractional Quantum Hall State

We have seen that the Laughlin wavefunction permits only odd-denominator FQHSs, so the observation of an *even* denominator state was unexpected. This even denominator state was discovered in 1988—the $\nu = 5/2$ fractional quantum Hall state [37]. It was conclusively shown to be a FQHS, with exactly quantized Hall plateau, in 1999 [38] (Fig. 1.9). The origin of this FQHS could not be immediately explained in the composite fermion formalism: indeed, at half-fillings, all flux quanta should be bound to electrons, and the composite fermions should experience zero effective magnetic field, as at $\nu = 1/2$ and $\nu = 3/2$.

Insight from the Bardeen–Cooper–Schrieffer theory of superconductivity provided a possible solution to the issue: pairing of composite fermions. It was realized that the composite fermions could pair up and condense like bosons into a new FQHS, analogous to the Cooper pairing of electrons in superconducting phase [39]. Haldane and Rezayi were among the first to propose a state formed of a sea of s-wave paired of composite fermions, yielding a spin-unpolarized FQHS [39]. While this particular proposed state was later discarded, it became clear that a FQHS of paired composite fermions was likely the best descriptor of $\nu = 5/2$ [40, 41].

Fig. 1.9 The $\nu = 5/2$ FQHS, with quantized Hall plateau and distinct R_{xx} minimum. This state breaks the odd-denominator FQHS formalism originated by Laughlin and Jain, and is expected to have non-Abelian properties. Ref. [38]. Reprinted figure with permission from W. Pan et al., Phys. Rev. Lett. 83, 3530 (1999). Copyright 1999 by the American Physical Society

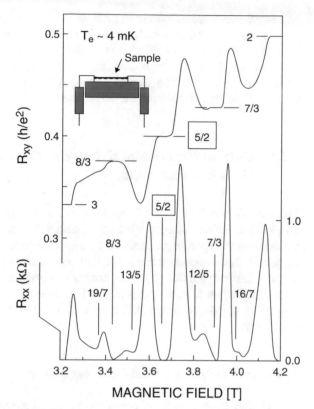

Moore and Read proposed a wavefunction to describe the paired ground state that had even more exciting implications: the Pfaffian wavefunction, also called the Moore–Read wavefunction [42]. This wavefunction is given by

$$\Psi_{MR} = \prod_{i<j}(r_i - r_j)^2 \mathrm{Pf}\left[\frac{1}{r_i - r_j}\right] \exp\left[-\frac{1}{4l_B}\sum_i |r_i|^2\right] \qquad (1.11)$$

This notably differs from the Laughlin wavefunction by the presence of the Pfaffian factor, $\mathrm{Pf}[\frac{1}{r_i-r_j}]$, which does the job of antisymmetrizing its argument, the positions of the electrons. It describes a ground state of p-wave paired composite fermions. Interestingly, this state would generate quasiparticles of non-Abelian statistics [42–44]. Under interchange of two quasiparticles, the system would be described by an entirely different ground state, unlike fermions, bosons, or anyons, which would only pick up an overall phase. Excitingly, these non-Abelian particles could host a platform for quantum computing according to certain proposals [44–46]. Morf [47] provided numerical work which provided strong evidence for several properties of the FQHS at $\nu = 5/2$, namely that it was indeed incompressible, spin polarized, and had strong overlap with the proposed Pfaffian state.

Several other models have been proposed to describe the $v = 5/2$ FQHS, notably the 331 model [48], which describes a paired state with *abelian* statistics, and the anti-Pfaffian state [49, 50], which is the particle-hole conjugate of the Pfaffian state, and which would also possess non-Abelian statistics. Also with non-Abelian statistics is the $U(1) \times SU_2(2)$ state proposed by Wen [51]. The goal of experiment is to try to distinguish between these proposed states.

1.6.1   Current Experimental Status of the $v = 5/2$ Fractional Quantum Hall State

The various proposed ground states of the $v = 5/2$ fractional quantum Hall state each have their own expected experimental signatures. Numerous experiments targeting these expected properties have therefore been undertaken. I briefly review some of the most important ones here. Current evidence tends to support the Pfaffian or anti-Pfaffian state, though direct, conclusive evidence for its non-Abelian nature remains to be found. Further experiment is needed to conclude the true nature of the $v = 5/2$ fractional quantum Hall state, and it therefore remains a subject of much excitement.

Gap of the $v = 5/2$ Fractional Quantum Hall State

One of the most persistent obstacles to characterizing $v = 5/2$ is that its experimentally measured excitation gap is apparently sample dependent, and nearly always smaller than predicted by numerical simulations, often by a factor of 20 [52–54]. It was proposed by Morf and D'Ambrumenil that the discrepancies were likely due to disorder, causing the gap to be reduced [54]. Evidence for the dependence of the gap on sample parameters was indeed seen experimentally [55–57]. This fact continues to thwart a consensus on the wavefunction describing the $v = 5/2$ FQHS.

Spin Polarization Studies

As discussed above, the Pfaffian and anti-Pfaffian ground states are fully spin polarized. Hence, a major experimental effort has been undertaken to determine the spin polarization state of $v = 5/2$. The predominant method for probing the spin polarization of a FQHS is by tilting a 2DES within a magnetic field. This procedure increases the Zeeman energy (dependent on the total magnetic field) while holding the system at fixed filling factor, as filling factor only depends on perpendicularly applied field. If the ground state is already polarized, the FQHS will simply become more robust as the spin energy increases. If the ground state is unpolarized, a *spin transition* occurs when the spin energy becomes equal to the cyclotron energy. At that point, a level crossing occurs, and the FQHS gap closes, meaning there is no signature of the FQHS in transport.

Eisenstein et al. [58] probed the $\nu = 5/2$ polarization in this manner. The $\nu = 5/2$ minimum steadily weakens and is destroyed, which was taken to be evidence of a spin-unpolarized state. However, it was then realized [59, 60] that the in-plane magnetic field was not merely destroying the quantum Hall state, it was in fact inducing an anisotropic state. This cast doubt on the idea that the destruction of the 5/2 quantum Hall state was caused by a spin transition, and necessitated further experiments in which rotational symmetry was not broken by external fields.

Gated samples were also used to tune the electron density. This allows one to observe the same filling factor at higher perpendicular magnetic field as one increases the density, without the complication of an in-plane magnetic field. Pan et al. used a heterojunction insulating gate field effect transistor (HIGFET) of GaAs and AlGaAs to deplete the 2DES in the GaAs while observing the $\nu = 5/2$ FQHS [61]. The authors concluded that the $\nu = 5/2$ FQHS exhibited behavior consistent with a spin polarized state. Later, Nuebler et al. studied a higher quality sample tuned by an in situ back gate from 0 to $2.7 \times 10^{11}\,\mathrm{cm}^{-2}$ [62]. This was a much smaller density range, but they likewise found only a monotonic increase of the gap with increasing density. This implied the $\nu = 5/2$ FQHS is spin polarized over this density range.

Other experimental methods pointed to full polarization of $\nu = 5/2$. Most prominent were nuclear magnetic resonance studies in which the Knight shift was probed [63]. The Knight shift is a study of the degree of polarization of an electron state. It is a shift in the resonance peak from that of bare arsenic atoms making up the quantum well due to the hyperfine interaction of electrons with these atoms. The larger the shift from the bare peak, the greater the polarization. By studying the Knight shift near filling factor 5/2, the researchers found evidence that the $\nu = 5/2$ FQHS is fully spin polarized.

While there is a substantial body of evidence that $\nu = 5/2$ is fully polarized, none of the experiments is perfectly conclusive. There also exist a handful of experiments that would seem to support that $\nu = 5/2$ is in fact unpolarized, mainly relying on the interaction of spinful electrons with polarized light [64, 65]. For this reason, the polarization status of the $\nu = 5/2$ FQHS has not been concluded.

Shot Noise and the Quasiparticle Charge

One exciting aspect of the FQHSs is the generation of the quasiparticle excitations with fractional charge and fractional statistics. For the Moore–Read Pfaffian, the quasiparticle charge is expected to be $e/4$ [42]. An elegant method for probing the charge of a current carrier is through the shot noise. Shot noise arises in a system where the charge carrier has tunneled through some kind of barrier—an electron emitted from a vacuum tube electrode, for example, or through a p–n junction barrier [28, 29]. The current that tunnels through the barrier has a component of its noise that obeys Poisson statistics. This component of the noise is called the shot noise, which depends on the charge of the emitted carriers.

To probe the shot noise in the FQHE, edge states of charge carriers were made to tunnel through a barrier induced by a quantum point contact (QPC). A QPC is a narrow constriction (on the order of a few hundred nm to a few μm) created by applying a voltage to narrow nanofabricated fingers patterned on the surface of a 2DES. The 2DES is depleted in this constriction, eventually to the point where the tunneling amplitude of the edge states through the constriction can be controlled. In this regime, the current noise is describable by a shot noise dependent on the charge of the carriers in the edge states that tunnel through the constriction.

Saminadayar et al. [66] performed this experiment at $\nu = 1/3$, finding charge carriers of $e^* = e/3$, as predicted by the theory of Laughlin [22]. de Picciotto et al. also obtained this result [67], demonstrating that indeed fractionally charged particles were generated at this state. Later on, Dolev et al. also studied the shot noise at the $\nu = 5/2$ FQHS [68]. They found a signature consistent with $e^* = e/4$, which is consistent with several of the proposed models for $\nu = 5/2$, but does not rule out any of them.

One more notable experiment was the detection of chargeless neutral modes, carrying only energy [69]. This is expected to be a Majorana mode, and lends evidence to the Pfaffian or anti-Pfaffian wavefunction as the ground states at $\nu = 5/2$.

Tunneling Conductance Through a Quantum Point Contact

Another parameter that may narrow down the possible ground states at $\nu = 5/2$ is called the interaction parameter, g. Radu et al. pioneered this work [70]. As in the case of the shot noise experiments, a quantum point contact was fabricated on the surface of their sample, depleting the 2DES in a narrow constriction. This time, the tunneling conductance through the QPC—effectively the I–V curve—was measured. The interaction parameter g was obtained by fitting the obtained tunneling conductance to an equation put forward by the model of Ref. [71] for weak tunneling of the quantum Hall edge states through the QPC. This g has a unique value, along with the quasiparticle charge, for each of the proposed theories. Radu et al. found that the best fit to their data was $g = 1/2$, with $e^* = e/4$, consistent with the wavefunctions that predict non-Abelian statistics, the Pfaffian and anti-Pfaffian [49, 50]. However, a subsequent similar experiment by Lin et al. [72] found that a better fit to the data was given by $g = 3/8$ and $e^* = e/4$. Such a result is consistent with the Abelian 331 state [48]. Therefore, ambiguity remains in the determination of the $\nu = 5/2$ ground state by this method.

Quantum Hall Interferometry

Fabry–Perot interferometry experiments have also attempted to illuminate the FQHS at $\nu = 5/2$. Willett [73–76] patterned two QPCs near one another onto a 2DES, making a quantum dot. Within this quantum dot, quantum Hall edge states could interfere with one another, making it effectively a Fabry–Perot interferometer. This interferometer could provide two tests of a quantum Hall state. First, it could give a measure of the quasiparticle charge. The edge states encircle magnetic flux through the quantum dot and demonstrate the Aharonov–Bohm effect. As the effective area of the quantum dot is changed by applying voltage to the QPCs, at a fixed magnetic field B, the resistance across the quantum dot oscillates with period h/e^*B. As in the shot noise experiment, Willett found the quasiparticle charge at $\nu = 5/2$ to be $e^* = e/4$ [73]. McClure et al. later performed a similar experiment at $\nu = 1/3, 2/3, 4/3$, and $5/3$, finding that $e^* = e/3$, helping to validate the ability of this experiment to detect charge.

The second possible utility of the quantum Hall interferometer is as a means of directly detecting non-Abelian statistics. The braiding properties could be probed in the interferometer, as applying voltage to the QPCs effectively directs the edge states in a path encircling other quasiparticles. If an even number of non-Abelian quasiparticles are encircled, Aharonov–Bohm oscillations should appear in the resistance, while the oscillations are suppressed if an odd number are encircled. Changing the area of the quantum dot by changing the QPC voltage should change the number of quasiparticles encircled, changing this oscillation signature. One would then expect in the non-Abelian case alternating patterns of resistance oscillations and absence of resistance oscillations. Willett [74] observed alternating patterns of period $e/4$ and $e/2$ when tuning the quantum dot area, and again saw this signature when holding the area fixed and sweeping magnetic field [75]. It was argued that this was still a signature of non-Abelian statistics, with the unexpected $e/2$ period oscillations coming from residual Abelian phases acquired [75]. Zhang et al. [77] showed that in small quantum dots, Coulomb interactions become important, and have a different resistance oscillation signature than the Aharonov–Bohm oscillations, emphasizing that care must be taken to ensure the quantum dot is not in the Coulomb blockade regime. While the interferometry results show promises of non-Abelian statistics, unequivocal results have not been acquired.

1.6.2 $\nu = 7/2$ Fractional Quantum Hall State

In the upper spin branch of the second Landau level, the states are weaker than in the lower spin branch, because they lie at lower magnetic field. Nonetheless, a FQHS exists at half-filled spin branch there as well: at $\nu = 7/2$ [78]. $\nu = 7/2$ is expected to be described by the same physics as at $\nu = 5/2$, though comes with a smaller activation energy, likely due to increased effects of disorder [54]. It is comparatively poorly studied because of its weakness, but in many cases provides additional evidence as to the nature of the mysterious half-filled states.

1.7   Conclusion

The quantum Hall effect, both integer and fractional, has been one of the most dramatic discoveries in solid state physics. Many questions remain as to the nature of certain fractional states, especially the $\nu = 5/2$ FQHS. For this reason, we are motivated to perform new experiments to discern its properties. Before I will describe my experimental technique, I will first describe another class of electronic ground states. The FQHSs, including that at $\nu = 5/2$, are isotropic states and can be described as topological ground states. In the next chapter, I will describe an important set of topologically trivial but spatially anisotropic phases: liquid crystalline states, which possess broken rotational and/or translational symmetries.

References

1. J.H. Davies, *The Physics of Low-Dimensional Semiconductors* (Cambridge University Press, Cambridge, 1998)
2. R.E. Prange, S.M. Girvin, *The Quantum Hall Effect* (Springer, New York, 1987)
3. J.K. Jain, *Composite Fermions* (Cambridge University Press, Cambridge, 2007)
4. J.P. Eisenstein, H.L. Stormer, Science **248**, 1510 (1990)
5. A. Stern, Ann. Phys. **323**, 204 (2008)
6. K. von Klitzing, G. Dorda, M. Pepper, Phys. Rev. Lett. **45**, 494 (1980)
7. D.C. Tsui, H.L. Stormer, A.C. Gossard, Phys. Rev. Lett. **48**, 1559 (1982)
8. K.S. Novoselov, A.K. Geim, S.V. Morozov, D. Jiang, Y. Zhang, S.V. Dubonos, I.V. Grigorieva, A.A. Firsov, Science **306**, 666 (2004)
9. K.S. Novoselov, A.K. Geim, S.V. Morozov, D. Jiang, M.I. Katsnelson, I.V. Grigorieva, S.V. Dubonos, A.A. Firsov, Nature **438**, 197 (2005)
10. A.H. Castro Neto, F. Guinea, N.M.R. Peres, K.S. Novoselov, A.K. Geim, Rev. Mod. Phys. **81**, 109 (2009)
11. K.I. Bolotin, F. Ghahari, M.D. Shulman, H.L. Stormer, P. Kim, Nature **462**, 196 (2009)
12. R. Williams, R.S. Crandall, A.H. Willis, Phys. Rev. Lett. **26**, 7 (1971)
13. A. Tsukazaki, S. Akasaka, K. Nakahara, Y. Ohno, H. Ohno, D. Maryenko, A. Ohtomo, M. Kawasaki, Nat. Mater. **9**, 889 (2010)
14. Q.H. Wang, K. Kalantar-Zadeh, A. Kis, J.N. Coleman, M.S. Strano, Nat. Nanotechnol. **7**, 699 (2012)
15. M.J. Manfra, Annu. Rev. Condens. Matter Phys. **5**, 347 (2014)
16. L.N. Pfeiffer, K.W. West, H.L. Stormer, K.W. Baldwin, Appl. Phys. Lett. **55**, 1888 (1989)
17. J.D. Jackson, *Classical Electrodynamics*, 2nd edn. (Wiley, New York, 1975)
18. R.F. Pierret, *Advanced Semiconductor Fundamentals*, 2nd edn. (Pearson Education, Upper Saddle River, 2003)
19. R. Shankar, *Principles of Quantum Mechanics* (Springer Science, New York, 1994)
20. B. Jeckelmann, B. Jeanneret, Séminaire Poincaré **2**, 39 (2004)
21. M. Büttiker, Phys. Rev. B **38**, 9375 (1988)
22. R.B. Laughlin, Phys. Rev. Lett. **50**, 1395 (1983)

23. F.D.M. Haldane, Phys. Rev. Lett. **51**, 605 (1983)
24. J.K. Jain, Phys. Rev. Lett. **63**, 199 (1989)
25. B.I. Halperin, Phys. Rev. Lett. **52**, 1583 (1984)
26. D. Arovas, J.R. Schrieffer, F. Wilczek, Phys. Rev. Lett. **53**, 722 (1984)
27. F. Wilczek, Phys. Rev. Lett. **49**, 957 (1982)
28. C. Beenaker, C. Schönenberger, Quantum shot noise. Phys. Today **56**, 37 (2003)
29. M. Reznikov, R. de Picciotto, M. Heiblum, D.C. Glattli, A. Kumar, L. Saminadayar, Superlattices and Microstruct. **23**, 901 (1998)
30. B.I. Halperin, P.A. Lee, N. Read, Phys. Rev. B **47**, 7312 (1993)
31. W. Kang, H.L. Stormer, L.N. Pfeiffer, K.W. Baldwin, K.W. West, Phys. Rev. Lett. **71**, 3850 (1993)
32. R.R. Du, H.L. Stormer, D.C. Tsui, L.N. Pfeiffer, K.W. West, Phys. Rev. Lett. **70**, 2944 (1993)
33. R.R. Du, H.L. Stormer, D.C. Tsui, A.S. Yeh, L.N. Pfeiffer, K.W. West, Phys. Rev. Lett. **73**, 3274 (1994)
34. R.L. Willett, R.R. Ruel, K.W. West, L.N. Pfeiffer, Phys. Rev. Lett. **71**, 3846 (1993)
35. M.Z. Hasan, C.L. Kane, Rev. Mod. Phys. **82**, 3045 (2010)
36. X.L. Qi, S.C. Zhang, Rev. Mod. Phys. **83**, 1057 (2011)
37. R.L. Willett, J.P. Eisenstein, H.L. Stormer, D.C. Tsui, A.C. Gossard, J.H. English, Phys. Rev. Lett. **59**, 1776 (1987)
38. W. Pan, J.-S. Xia, V. Shvarts, D.E. Adams, H.L. Stormer, D.C. Tsui, L.N. Pfeiffer, K.W. Baldwin, K.W. West, Phys. Rev. Lett. **83**, 3530 (1999)
39. F.D.M. Haldane, E.H. Rezayi, Phys. Rev. Lett. **60**, 956 (1988)
40. N. Read, D. Green, Phys. Rev. B **61**, 10267 (2000)
41. K. Park, V. Melik-Alaverdian, N.E. Bonesteel, J.K. Jain, Phys. Rev. B **58**, 10167 (1998)
42. G. Moore, N. Read, Nucl. Phys. B **360**, 362 (1991)
43. M. Greiter, X.G. Wen, F. Wilczek, Phys. Rev. Lett. **66**, 3205 (1991)
44. C. Nayak, S.H. Simon, A. Stern, M. Freedman, S. Das Sarma, Rev. Mod. Phys. **80**, 1083 (2008)
45. S. Das Sarma, M. Freedman, C. Nayak, Phys. Rev. Lett. **94**, 166802 (2005)
46. A.Y. Kitaev, Ann. Phys. **303**, 2 (2003)
47. R.H. Morf, Phys. Rev. Lett. **80**, 1505 (1998)
48. B.I. Halperin, Helv. Phys. Acta **56**, 75 (1983)
49. M. Levin, B.I. Halperin, B. Rosenow, Phys. Rev. Lett. **99**, 236806 (2007)
50. S.S. Lee, S. Ryu, C. Nayak, M.P. Fisher, Phys. Rev. Lett. **99**, 236807 (2007)
51. X.G. Wen, Phys. Rev. Lett. **66**, 802 (1991)
52. A.E. Feiguin, E. Rezayi, C. Nayak, S. Das Sarma, Phys. Rev. Lett. **100**, 166803 (2008)
53. R.H. Morf, N. d'Ambrumenil, S. Das Sarma, Phys. Rev. B **66**, 075408 (2002)
54. R.H. Morf, N. d'Ambrumenil, Phys. Rev. B **68**, 113309 (2003)
55. C.R. Dean, B.A. Piot, P. Hayden, S. Das Sarma, G. Gervais, L.N. Pfeiffer, K.W. West, Phys. Rev. Lett. **100**, 146803 (2008)
56. C.R. Dean, B.A. Piot, P. Hayden, S. Das Sarma, G. Gervais, L.N. Pfeiffer, K.W. West, Phys. Rev. Lett. **101**, 186806 (2008)
57. N. Samkharadze, J.D. Watson, G.C. Gardner, M.J. Manfra, L.N. Pfeiffer, K.W. West, G.A. Csáthy. Phys. Rev. B **84**, 121305 (2011)
58. J.P. Eisenstein, R. Willett, H.L. Stormer, D.C. Tsui, A.C. Gossard, J.H. English, Phys. Rev. Lett. **61**, 997 (1988)
59. W. Pan, R.R. Du, H.L. Stormer, D.C. Tsui, L.N. Pfeiffer, K.W. Baldwin, K.W. West, Phys. Rev. Lett. **83**, 820 (1999)
60. M.P. Lilly, K.B. Cooper, J.P. Eisenstein, L.N. Pfeiffer, K.W. West, Phys. Rev. Lett. **83**, 824 (1999)
61. W. Pan, H.L. Stormer, D.C. Tsui, L.N. Pfeiffer, K.W. Baldwin, K.W. West, Solid State Commun. **119**, 641 (2001)
62. J. Nuebler, V. Umansky, R. Morf, M. Heiblum, K. von Klitzing, and J. Smet. Phys. Rev. B **81**, 035316 (2010).
63. L. Tiemann, G. Gamez, N. Kumada, K. Muraki, Science **335**, 828 (2012)

64. M. Stern, P. Plochocka, V. Umansky, D.K. Maude, M. Potemski, I. Bar-Joseph, Phys. Rev. Lett. **105**, 096801 (2010)
65. T.D. Rhone, J. Yan, Y. Gallais, A. Pinczuk, L.N. Pfeiffer, K.W. West, Phys. Rev. Lett. **106**, 196805 (2011)
66. L. Saminadayar, D.C. Glattli, Y. Jin, B. Etienne, Phys. Rev. Lett. **79**, 2526 (1997)
67. R. de Picciotto, M. Reznikov, M. Heiblum, V. Umansky, G. Bunin, D. Mahalu, Nature **389**, 162 (1997)
68. M. Dolev, M. Heiblum, V. Umansky, A. Stern, D. Mahalu, Nature **452**, 829 (2008)
69. A. Bid, N. Ofek, H. Inoue, M. Heiblum, C.L. Kane, V. Umansky, D. Mahalu, Nature **466**, 585 (2010)
70. I.P. Radu, J.B. Miller, C.M. Marcus, M.A. Kastner, L.N. Pfeiffer, K.W. West, Science **320**, 899 (2008)
71. X.G. Wen, Phys. Rev. B **44**, 5708 (1991)
72. X. Lin, C. Dillard, M.A. Kastner, L.N. Pfeiffer, K.W. West, Phys. Rev. B **85**, 165321 (2012)
73. R.L. Willett, L.N. Pfeiffer, K.W. West, Proc. Natl. Acad. Sci. **106**, 8853 (2009)
74. R.L. Willett, L.N. Pfeiffer, K.W. West, Phys. Rev. B **82**, 205301 (2010)
75. R.L. Willett, C. Nayak, K. Shtengel, L.N. Pfeiffer, K.W. West, Phys. Rev. Lett. **111**, 186401 (2013)
76. R.L. Willett, Rep. Prog. Phys. **76**, 076501 (2013)
77. Y. Zhang, D.T. McClure, E.M. Levenson-Falk, C.M. Marcus, L.N. Pfeiffer, K.W. West, Phys. Rev. B **79**, 241304 (2009)
78. J.P. Eisenstein, K.B. Cooper, L.N. Pfeiffer, K.W. West, Phys. Rev. Lett. **88**, 076801 (2002)

Chapter 2
The Quantum Hall Nematic Phase

The quantum Hall states are not the only possible electron phases in the two-dimensional electron system in a strong magnetic field. The 2DES may also host various crystalline phases, such as the Wigner crystal [1, 2], the bubble phases (also known as the reentrant integer quantum Hall states) [3–5], and the electron nematic phase, commonly referred to as the stripe phase[6–10]. In contrast to the quantum Hall states which are topological phases, these phases are traditional Landau phases with charge order. The nematic phase breaks rotational symmetry, and is characterized by highly anisotropic longitudinal resistance. Importantly, the nematic phase is the ground state at half-filling in the third and higher Landau levels, at $\nu = 9/2, 11/2, 13/2$, and so on, marking yet another departure from the composite fermion picture at half-fillings. I will focus my discussion on the nematic phase.

2.1 Nematicity in Condensed Matter Systems

Nematic phases are ubiquitous in material systems. In liquid crystal systems, long, chain-like molecules can arrange themselves in ways that break spatial symmetries [11, 12]. In the nematic phase, the molecules arrange themselves end to end in long chains while retaining the properties of a fluid, namely freedom of the molecules to move with respect to one another. Rotational symmetry is broken in this nematic phase, but translational symmetry is preserved, as the molecules do not form periodic arrays. The ability to drive a phase transition from an isotropic to a nematic phase in liquid crystal systems underlies many liquid crystal displays in modern electronics.

The possibility of nematic order arising in a solid state system was made real by consideration of cuprate high temperature superconductors [13]. The superconducting phase in these materials emerges by doping a highly insulating

© Springer Nature Switzerland AG 2019
K. A. Schreiber, *Ground States of the Two-Dimensional Electron System at Half-Filling under Hydrostatic Pressure*, Springer Theses,
https://doi.org/10.1007/978-3-030-26322-5_2

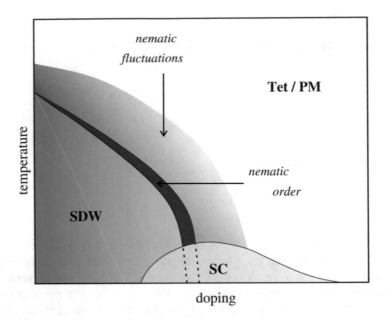

Fig. 2.1 An example of a schematic phase diagram of an iron pnictide high-T_c superconductor. Nematic order can be seen above the superconducting region (yellow, labeled SC). Nematic order may be important to pairing correlations in the superconducting phase. The white region labeled Tet/PM denotes a paramagnet phase, and the blue region labeled SDW denotes a spin density wave phase. From Ref. [20]. R.M. Fernandes and J. Schmalian "Manifestations of nematic degrees of freedom in the magnetic, elastic, and superconducting properties of the iron pnictides" Supercond. Sci. Technol., 25, 084005 (2012). ©IOP Publishing. Reproduced with permission. All rights reserved

antiferromagnetic phase [14, 15]. It was realized that striped charge order could be an intermediate phase between this insulating phase and the superconducting phase. Further theoretical progress was made by allowing the stripe to fluctuate in time— giving it the properties of an electronic nematic liquid crystal [15–17]. Signs of such a nematic order have indeed been experimentally detected [14, 18]. Nematic order is also thought to play an important role in iron pnictide high-T_c superconductors [19, 20], which is demonstrated in a schematic phase diagram in Fig. 2.1.

Signatures of nematic order have now been observed in a variety of condensed matter systems. Beyond cuprate and pnictide high-T_c superconductors, even more unusual superconductors such as the correlated electron oxide material $Sr_3Ru_2O_7$ [21] display nematics. Topological materials such as bismuth [22] and certain cerium based heavy fermion materials [23] also display nematic-like anisotropy in the presence of symmetry-breaking magnetic fields. Finally, the nematic phase is seen in the GaAs 2DES [6, 7], on which I will focus most of this chapter.

The exact natures of these nematic phases are not all perfectly identified. The most common picture is that the nematic is a melted charge density wave. However, other possibilities could be a spin density wave, or in the case of high-T_c superconductors, a pair density wave, modulating between regions of paired electrons and normal regions [14]. In what follows, I will explore the nematic phase in the 2DES.

2.2   Prediction and Theory of the Nematic State in the Two-Dimensional Electron System

In 1996, Fogler, Koulakov, and Shklovskii theoretically studied the electron ground states at high filling factor, motivated by the fact that as yet, no FQHSs had ever been observed at filling factors greater than 4 [8, 24] in the GaAs 2DES. This was surprising, since electron interactions are expected to play just as significant a role at these filling factors as in the lower Landau levels where the FQHE is observed. Upon analysis of the problem through Hartree–Fock approximation, which is a mean-field theory, they predicted the existence of charge density waves (CDWs) in the third and higher Landau levels. These CDWs feature areas of modulated charge density which break spatial symmetries, and were referred to as the stripe phases and bubble phases. The modulation of charge in this regime of filling factors has its roots in the fact that the Landau level wavefunction has nodes for high Landau level index. The stripe phase in particular was predicted to appear at half-filling in the third and higher Landau levels. The stripe phase breaks rotational symmetry in that the electrons form a periodic array of stripes in the plane of the sample. The theoretical prediction of the stripelike CDW was corroborated by Moessner and Chalker [9].

Work on electron liquid phases in general condensed matter systems also supported the formation of a broken rotational symmetry phase at half-fillings. Fradkin, Kivelson et al. elucidated on the CDW picture by considering it at finite temperature [10, 25]. Rather than a static CDW of fixed stripes, they permitted stripes to fluctuate in time, like a liquid crystal would. They found that a truly periodic liquid crystal, called the smectic, which breaks both rotational and translational symmetry, exists only for extremely low temperature and disorder. At higher temperature and disorder, the nematic phase is favored, which breaks rotational but not translational symmetry. The smectic and the nematic are pictured in Fig. 2.2.

2.3   Experimental Observation of the Nematic Phase: $v = 9/2, 11/2, 13/2...$

The predictions of the theorists proved to be fruitful in 1999. The nematic phase was first observed in the quantum Hall regime in GaAs by two groups: Lilly et al. [6] and Du et al. [7] in the third and higher Landau levels, at $v = 9/2, 11/2, 13/2$, and so on. It revealed itself as an extremely high longitudinal resistance when the current was passed and the voltage was measured along the $\langle 1\bar{1}0 \rangle$ crystal lattice direction ("hard" axis, resistance denoted R_{xx}) and an extremely low resistance when measured along the $\langle 110 \rangle$ direction ("easy" axis, resistance denoted R_{yy})

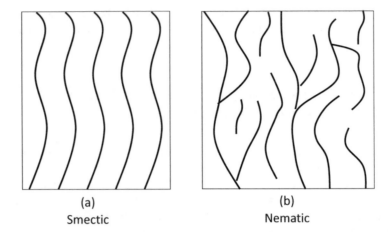

| (a) | (b) |
| Smectic | Nematic |

Fig. 2.2 (**a**) The periodic smectic phase. This phase is a liquid crystal that breaks translational and rotational symmetry, and is thought to exist at extremely low temperature and disorder. (**b**) The nematic phase, at finite temperature and disorder. This is a liquid crystal that breaks rotational symmetry while preserving translational symmetry

(Fig. 2.3). This was consistent with the stripelike features formed in the nematic phase as in Fig. 2.2. When current is passed along the stripelike features, the very low resistance is measured, but for current to flow *across* the stripelike features, electrons suffer a large amount of scattering, so a high resistance is measured. The anisotropy was most pronounced at low temperatures, and quickly diminished with increasing temperature. This was exciting confirmation of other exotic electron ground states beyond the quantum Hall effect that could arise in these 2D electron systems. The nematic state at $\nu \geq 9/2$ at half-filling is now routinely observed in high quality samples.

An important aspect of the nematic phase is that it arises in purely perpendicular magnetic field, without any externally applied symmetry-breaking force. In this sense, we say that the original nematic phase at half-filling in the third and higher Landau levels is a spontaneously broken symmetry phase. It should be emphasized that in typical samples in pure perpendicular field, $\nu = 5/2$ and $\nu = 7/2$ are isotropic fractional quantum Hall states, and likewise $\nu = 1/2$ and $\nu = 3/2$ are isotropic composite Fermi seas.

2.4   The Effect of In-Plane Magnetic Field on the Nematic at $\nu = 9/2, 11/2, 13/2...$

Soon after the nematic phase was discovered, studies were completed tilting the system in a magnetic field [26, 27]. The in-plane field $B_\parallel$ had some surprising effects. First, the nematic phase at $\nu = 9/2, 11/2, 13/2...$ was modified. Most

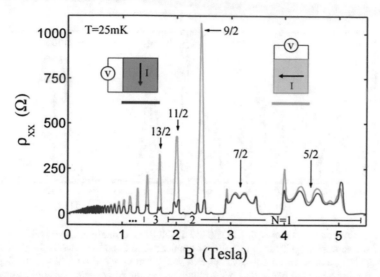

Fig. 2.3 The nematic phase, which was originally discovered by Lilly et al. [6] and Du et al. [7] in 1999. The huge resistance anisotropy can be clearly seen at $\nu = 9/2$, $11/2$, $13/2$, and so on. The green trace is measured along the $\langle 1\bar{1}0 \rangle$ crystallographic direction of GaAs, and the red trace is measured along the $\langle 110 \rangle$. The stripelike features formed by the electrons are aligned with $\langle 110 \rangle$. Note that $\nu = 7/2$ and $\nu = 5/2$ are isotropic. Ref. [6]. Reprinted figure with permission from M.P. Lilly et al. [6]. Copyright 1999 by the American Physical Society

strikingly, when $B_\parallel$ was applied along $\langle 110 \rangle$—the easy axis—as the tilt angle and thus the magnitude of the in-plane field increased, the system became isotropic and then anisotropic again, but with *switched* hard and easy axes. That is, one measured high resistance along the $\langle 110 \rangle$ crystal direction and low resistance along $\langle 1\bar{1}0 \rangle$. When $B_\parallel$ was applied along $\langle 1\bar{1}0 \rangle$, the nematic was not strongly affected, showing only an eventual reduction in anisotropy at $\nu = 9/2$. This suggested that the in-plane field played a strong role in re-orienting the direction of the stripes.

2.5   The Effect of In-Plane Magnetic Field on the Second Landau Level Fractional Quantum Hall States

The second notable effect of in-plane field was that dramatic anisotropy set in at $\nu = 5/2$ and $\nu = 7/2$ [26, 27]. In-plane field was applied along the $\langle 110 \rangle$ and the $\langle 1\bar{1}0 \rangle$ crystalline directions. The direction of the in-plane field set the direction along which the resistance peak was measured for $\nu = 5/2$ and $\nu = 7/2$. It was further found that in certain samples, one could induce anisotropy at $\nu = 5/2$, and then at even higher tilt, return it to an isotropic state [28]. The anisotropic state induced by in-plane field at $\nu = 5/2$ and $\nu = 7/2$ is also referred to as the nematic phase, like that in the third Landau level. It is believed that this induced nematic phase in the

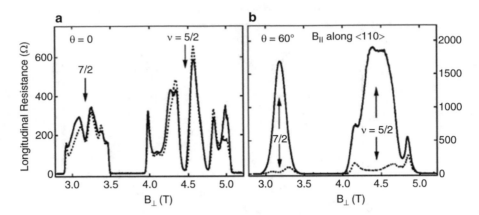

Fig. 2.4 (a) At zero tilt, the $\nu = 5/2$ FQHS has a well-defined minimum, and both $\nu = 5/2$ and $\nu = 7/2$ are isotropic. (b) Tilting the 2DES in magnetic field to $60°$, so that the in-plane field $B_{\parallel}$ lies along $\langle 110 \rangle$, a huge anisotropy develops at $\nu = 5/2$ and $\nu = 7/2$, as well surrounding filling factors. The in-plane field explicitly breaks rotational symmetry, inducing this anisotropy across the second Landau level. Ref. [26]. Reprinted figure with permission from M.P. Lilly et al. [26]. Copyright 1999 by the American Physical Society

second Landau level is indeed related to the spontaneously arising nematic phase in the third Landau level. The key difference is that the in-plane field here breaks the rotational symmetry in the $x - y$ plane of the system explicitly in the Hamiltonian, while no such explicit rotational symmetry breaking terms arise in the Hamiltonian when only a perpendicular magnetic field is applied. The transport signature of the induced nematic is also markedly different from that of the spontaneous nematic. With the application of in-plane field, anisotropy arises throughout the entire second Landau level, as seen in Fig. 2.4. The spontaneously arising nematic phase, when no in-plane field is applied, however, is limited to a much narrower range of filling factor around half-filling.

2.5.1   Nematic Fractional Quantum Hall States: $\nu = 7/3$ and $\nu = 5/2$

Tilted-field experiments have also revealed emergent *incompressible* anisotropic states. Experiments at $\nu = 7/3$ have revealed a quantized Hall plateau coexistent with anisotropic longitudinal resistance. This effect strengthened with increasing in-plane field, but diminished at very high in-plane field [29]. It has been proposed that the development of this anisotropy is a signature of a phase transition to a nematic fractional quantum Hall state [30]. Similar results have been seen at $\nu = 5/2$ as the sample is tilted in magnetic field [31], which the authors suggested is reminiscent of a nematic FQHS.

2.6   Recent Studies of the Nematic Phase

The question of why the nematic phase always orients along the same crystalline directions, in almost every GaAs sample measured, is one that is still actively studied. The $\langle 1\bar{1}0 \rangle$ and $\langle 110 \rangle$ are symmetric in GaAs, so electric fields of the host heterostructure may play a role. Indeed, Pollanen et al. [32] systematically studied several different heterostructure types, finding little dependence of the hard and easy axes on the structure for quantum well samples. In single heterojunction samples, however, the depth of the 2DES beneath the surface played a role in determining the nematic phases orientation. Relatedly, Shi et al. examined how the dopant setback layer affects the anisotropy when in-plane magnetic fields are applied [33].

Recently, numerous experiments have been done studying how other experimental parameters may reorient the direction of the nematic. Shi et al. have done extensive work considering how temperature, electron density, and tilt angle affect the hard and easy axis directions [34–37]. Göres et al. were able to enhance or diminish anisotropy by selectively applying a DC bias along the two crystalline directions in addition to the excitation current used to probe the resistance [38]. Mueed et al. fabricated a periodic strain grating using e-beam resist along the two perpendicular crystalline directions and were able to reorient the anisotropy directions [39]. The potential induced by this grating was strong enough to compete significantly with in-plane magnetic field in selecting the anisotropy axes. Liu et al. observed the behavior of anisotropic phases in wide (around 60 nm) quantum wells [40].

Other notable experiments probe the microscopic structure of the nematic phase. The group of Smet [41, 42] applied surface acoustic waves, finding evidence for a periodicity of the stripelike structures in the nematic, and found evidence for negative permittivity in the nematic and bubble phases. Msall and Dietsche [43] also use acoustic techniques to probe the stripes. Sambandamurthy et al. [44] find evidence for a pinning mode of the stripes, observing a radio-frequency resonance when the signal is applied along the hard axis of the nematic. Finally, Qian et al. [45] observed a temperature-driven transition from the nematic phase to the smectic phase. The nematic phase, at the time of this writing, is not observed in the quantum Hall regime of any other 2D materials except GaAs, suggesting that the internal crystalline fields of GaAs are uniquely strong enough to stabilize the stripelike structures.

Recent theoretical efforts towards understanding the nematic phase have also been made. Many of these probe the possible instabilities of the nematic phase toward other phases, especially the composite Fermi sea and the paired FQHSs, following in the footsteps of Haldane and Rezayi, who predicted a transition between a stripe phase and a paired FQHS [49]. I will discuss the implication of these recent theories later in this thesis, as they pertain to the pressure-driven FQHS-to-nematic transition that we observe.

Fig. 2.5 Weak anisotropy
arises at $\nu = 7/2$ in a very
low density sample,
$n = 5 \times 10^{10}\,\mathrm{cm}^{-2}$. Ref.
[52]. Reprinted figure with
permission from W. Pan et al.
[52]. Copyright 2014 by the
American Physical Society

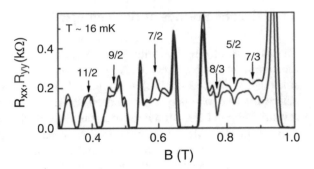

2.7   Other Anisotropic Signatures in Even Denominator States

Nematic phases have been observed in a handful of other instances in the quantum Hall regime. Uniaxial strain induces anisotropy at $\nu = 5/2$ and $\nu = 7/2$, due to the enhanced crystalline electric fields [50]. Spontaneously arising anisotropy was observed in a two-dimensional hole system at $\nu = 7/2$, where the effect was attributed to strong spin–orbit coupling [51].

Interestingly, Pan et al. observed a weak anisotropic phase at $\nu = 7/2$ in a two-dimensional electronic system, without any applied in-plane magnetic field [52]. This weak anisotropy is presented in Fig. 2.5. This spontaneously arising anisotropy may represent a nascent nematic phase. Importantly, this sample was of a very low density, $n = 5 \times 10^{10}\,\mathrm{cm}^{-2}$. Such an unusual phase at $\nu = 7/2$ may be an important result of physics in a low electron density sample, to which I will return later after discussing my experimental results.

2.8   Electron Solids: Wigner Crystal and Bubble Phases

For completeness I will here review the two other major types of compressible electronic phases possessing broken spatial symmetry: the Wigner crystal and the bubble phases. The Wigner crystal is found at very high magnetic field and correspondingly very low filling factor, $\nu < 1/5$ [1, 2]. In this limit, the cyclotron radius and the kinetic energy of the electrons are very small, effectively localizing the electrons. The electrons become pinned in a periodic array expected to be a triangular lattice. This leads to an insulating state, manifesting in a huge resistance at the highest magnetic fields.

The reentrant integer quantum Hall states, also known as the bubble phases, are also a type of electron solid [4, 5, 24]. Like the stripe phase, they are a charge density wave, and they occur in the flanks of Landau level spin branches, unlike the stripe phase which occurs around half-filling. The bubble phases occur beginning in the *second* Landau level in a typical sample, and persist into the third and higher Landau

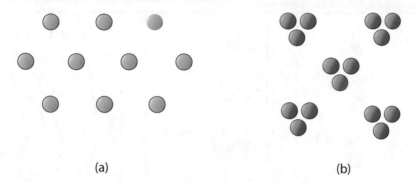

(a)   (b)

Fig. 2.6 (a) A cartoon of the Wigner crystal, a highly insulating phase of localized electrons in a crystalline formation. (b) The bubble phase made up of small clusters, or bubbles, of electrons. The bubbles themselves are localized in a crystal as well

levels. They are most robust in the second Landau level, where there are four in each spin branch, and become weaker in the third and higher Landau levels, which possess only two per spin branch. Their signature in transport is an isotropic minimum in the longitudinal resistance, and a plateau in the Hall resistance quantized to the nearest *integer* value of h/e^2. They are highly insulating phases which do not contribute to conduction; hence, we measure integer conductance arising from the edge states of the nearest Landau level. They are expected to be triangular lattices of small clusters—that is, bubbles—of electrons forming a periodic array. Detailed studies of the temperature dependence of the bubble phases have recently been completed, demonstrating the collective nature of these states [53, 54]. Figure 2.6 shows schematic cartoons of the Wigner crystal and the bubble phase.

2.9   Summary of States at Half-Filling

It is useful to summarize here the results so far about the ground states at half-fillings in two-dimensional electron systems in purely perpendicularly applied fields (Fig. 2.7). In the lowest Landau level, at $\nu = 1/2$ and $\nu = 3/2$, we have the composite fermi sea, which fits neatly into the elementary composite fermion formalism. More recently, interest was renewed in this state due to the proposal that it could host Dirac fermions [55]. In the second Landau level, we have a FQHS at $\nu = 5/2$ and $\nu = 7/2$. This is a ground state of paired composite fermions. Notably, it is a topological state due to the existence of robust edge states [56, 57]. A phase transition from a FQHS involves the changing of topological order—that is, changing the number and type of edge states [56, 57]. In the third and higher Landau levels, at $\nu = 9/2, 11/2$, and so on, we have the nematic phase. It should be emphasized that the nematic phase is quite different from the states in the lower Landau levels: it is a traditional Landau phase exhibiting broken rotational

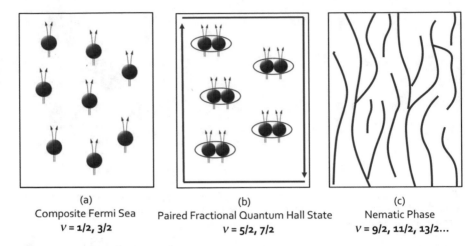

(a)   (b)   (c)
Composite Fermi Sea   Paired Fractional Quantum Hall State   Nematic Phase
$v = 1/2, 3/2$ $v = 5/2, 7/2$ $v = 9/2, 11/2, 13/2...$

Fig. 2.7 Schematic representation of the states at half-filling. (**a**) In the lowest Landau level, a composite fermion sea exists at $v = 1/2$ and $v = 3/2$. (**b**) In the second Landau level, a paired FQHS exists at $v = 5/2$ and $v = 7/2$. (**c**) In the third and higher Landau levels, at $v = 9/2, 11/2, 13/2...$ we have the nematic phase

symmetry [58]. Phase transitions of the nematic involve the changing of the nematic order parameter, just as in the case of melting of an ordered crystal [25, 58]. In contrast, phase transitions from the isotropic FQHSs do not involve the changing of such an order parameter.

The differences in the states at half-fillings in the different Landau levels likely have their roots in the different Landau level wavefunctions. An important question is therefore whether a sample could be tuned such that one half-filled Landau level could host a ground state usually found in *another* Landau level. This is a topic I will explore in the later chapters of this thesis.

2.10   Conclusion

The nematic phase is an important member of the wide variety of 2D electronic ground states. This state is a paradigm for spontaneously broken rotational symmetry. A matter of even more richness and depth is the question of whether and how a fractional quantum Hall state could have a spontaneously driven *transition* to the nematic phase. We in fact observe this, as I will discuss below, so an understanding of the nematic phase is necessary in order to consider this result. In the next chapter, I will describe how we drove this transition through the application of hydrostatic pressure.

References

1. W. Pan, H.L. Stormer, D.C. Tsui, L.N. Pfeiffer, K.W. Baldwin, K.W. West, Phys. Rev. Lett. **88**, 176802 (2002)
2. W. Pan, G.A. Csáthy, D.C. Tsui, L.N. Pfeiffer, K.W. West, Phys. Rev. B **71**, 035302 (2005)
3. J.S. Xia, W. Pan, C.L. Vicente, E.D. Adams, N.S. Sullivan, H.L. Stormer, D.C. Tsui, L.N. Pfeiffer, K.W. Baldwin, K.W. West, Phys. Rev. Lett. **93**, 176809 (2004)
4. K.B. Cooper, M.P. Lilly, J.P. Eisenstein, L.N. Pfeiffer, K.W. West, Phys. Rev. B **60**, 11285 (1999)
5. J.P. Eisenstein, K.B. Cooper, L.N. Pfeiffer, K.W. West, Phys. Rev. Lett. **88**, 076801 (2002)
6. M.P. Lilly, K.B. Cooper, J.P. Eisenstein, L.N. Pfeiffer, K.W. West, Phys. Rev. Lett. **82**, 394 (1999)
7. R.R. Du, D.C. Tsui, H.L. Stormer, L.N. Pfeiffer, K.W. Baldwin, K.W. West, Solid State Commun. **109**, 389 (1999)
8. A.A. Koulakov, M.M. Fogler, B.I. Shklovskii, Phys. Rev. Lett. **76**, 499 (1996)
9. R. Moessner, J.T. Chalker, Phys. Rev. B. **54**, 5006 (1996)
10. E. Fradkin, S.A. Kivelson, Phys. Rev. B. **59**, 8065 (1999)
11. I.-C. Khoo, *Liquid Crystals* (Wiley, Hoboken, 2007)
12. S. Chandrasekhar, Phys. Today **46**, 122 (1993)
13. J.G. Bednorz, K.A. Muller, Z. Phys. B **64**, 189 (1986)
14. B. Keimer, S.A. Kivelson, M.R. Norman, S. Uchida, J. Zaanen, Nature **518**, 179 (2015)
15. E. Fradkin, S.A. Kivelson, J.M. Tranquada, Rev. Mod. Phys. **87**, 457 (2015)
16. S.A. Kivelson, E. Fradkin, V.J. Emery, Nature **393**, 550 (1998)
17. V.J. Emery, S.A. Kivelson, J.M. Tranquada, Proc. Natl. Acad. Sci. **96**, 8814 (1999)
18. S.-W. Cheong, G. Aeppli, T.E. Mason, H. Mook, S.M. Hayden, P.C. Canfield, Z. Fisk, K.N. Clausen, J.L. Martinez, Phys. Rev. Lett. **67**, 1791 (1991)
19. R.M. Fernandes, A.V. Chubukov, J. Schmalian, Nat. Phys. **10**, 97 (2014)
20. R.M. Fernandes, J. Schmalian, Supercond. Sci. Technol. **25**, 084005 (2012)
21. R.A. Borzi, S.A. Grigera, J. Farrell, R.S. Perry, S.J.S. Lister, S.L. Lee, D.A. Tennant, Y. Maeno, A.P. Mackenzie, Science **315**, 214 (2007)
22. B.E. Feldman, M.T. Randeria, A. Gyenis, F. Wu, H. Ji, R.J. Cava, A.H. MacDonald, A. Yazdani, Science **354**, 316 (2016)
23. F. Ronning, T. Helm, K.R. Shirer, M.D. Bachmann, L. Balicas, M.K. Chan, B.J. Ramshaw, R.D. McDonald, F.F. Balakirev, M. Jaime, E.D. Bauer, P.J.W. Moll, Nature **548**, 313 (2017)
24. M.M. Fogler, Stripe and bubble phases in quantum hall systems, in *High Magnetic Fields*. Lecture Notes in Physics, vol. 595 (Springer, Berlin, 2002), pp. 98–138
25. E. Fradkin, S.A. Kivelson, M.J. Lawler, J.P. Eisenstein, A.P. Mackenzie, Annu Rev. Condens. Matter Phys. **1**, 153 (2010)
26. M.P. Lilly, K.B. Cooper, J.P. Eisenstein, L.N. Pfeiffer, K.W. West, Phys. Rev. Lett. **83**, 824 (1999)
27. W. Pan, R.R. Du, H.L. Stormer, D.C. Tsui, L.N. Pfeiffer, K.W. Baldwin, K.W. West, Phys. Rev. Lett. **83**, 820 (1999)
28. J. Xia, V. Cvicek, J.P. Eisenstein, L.N. Pfeiffer, K.W. West, Phys. Rev. Lett. **105**, 176807 (2010)
29. J. Xia, J.P. Eisenstein, L.N. Pfeiffer, K.W. West, Nat. Phys. **7**, 845 (2011)
30. M. Mulligan, C. Nayak, S. Kachru, Phys. Rev. B **84**, 195124 (2011)
31. Y. Liu, S. Hasdemir, M. Shayegan, L.N. Pfeiffer, K.W. West, K.W. Baldwin, Phys. Rev. B **88**, 035307 (2013)
32. J. Pollanen, K.B. Cooper, S. Brandsen, J.P. Eisenstein, L.N. Pfeiffer, K.W. West, Phys. Rev. B **92**, 115410 (2015)
33. X. Shi, W. Pan, K.W. Baldwin, K.W. West, L.N. Pfeiffer, D.C. Tsui, Phys. Rev. B **91**, 125308 (2015)
34. Q. Shi, M.A. Zudov, J.D. Watson, G.C. Gardner, M.J. Manfra, Phys. Rev. B **93**, 121404 (2016)
35. Q. Shi, M.A. Zudov, J.D. Watson, G.C. Gardner, M.J. Manfra, Phys. Rev. B **93**, 121411 (2016)

36. Q. Shi, M.A. Zudov, Q. Qian, J.D. Watson, M.J. Manfra, Phys. Rev. B. **95**, 161303 (2017)
37. Q. Shi, M.A. Zudov, B. Friess, J.H. Smet, J.D. Watson, G.C. Gardner, M.J. Manfra, Phys. Rev. B **95**, 161404 (2017)
38. J. Göres, G. Gamez, J.H. Smet, L. Pfeiffer, K. West, A. Yacoby, V. Umansky, K. von Klitzing, Phys. Rev. Lett. **99**, 246402 (2007)
39. M.A. Mueed, Md. Shafayat Hossain, L.N. Pfeiffer, K.W. West, K.W. Baldwin, M. Shayegan, Phys. Rev. Lett. **117**, 076803 (2016)
40. Y. Liu, D. Kamburov, M. Shayegan, L.N. Pfeiffer, K.W. West, K.W. Baldwin, Phys. Rev. B **87**, 075314 (2013)
41. B. Friess, Y. Peng, B. Rosenow, F. von Oppen, V. Umansky, K. von Klitzing, J.H. Smet, Nat. Phys. **13**, 1124 (2017)
42. B. Friess, V. Umansky, L. Tiemann, K. von Klitzing, J.H. Smet, Phys. Rev. Lett. **113**, 076803 (2014)
43. M.E. Msall, W. Dietsche, New J. Phys. **17**, 043042 (2015)
44. G. Sambandamurthy, R.M. Lewis, H. Zhu, Y.P. Chen, L.W. Engel, D.C. Tsui, L.N. Pfeiffer, K.W. West, Phys. Rev. Lett. **100**, 256801 (2008)
45. Q. Qian, J. Nakamura, S. Fallahi, G.C. Gardner, M.J. Manfra, Nat. Commun. **8**, 1536 (2017)
46. S. Basak, E.W. Carlson, Phys. Rev. B **96**, 081303 (2017)
47. Y. You, G.Y. Cho, E. Fradkin, Phys. Rev. B **93**, 205401 (2016)
48. A. Mesaros, M.J. Lawler, E.-A. Kim, Phys. Rev. B **95**, 125127 (2017)
49. E.H. Rezayi, F.D.M. Haldane, Phys. Rev. Lett. **84**, 4685 (2000)
50. S.P. Koduvayur, Y. Lyanda-Geller, S. Khlebnikov, G.A. Csáthy, M.J. Manfra, L.N. Pfeiffer, K.W. West, L.P. Rokhinson, Phys. Rev. Lett. **106**, 016804 (2011)
51. M.J. Manfra, R. de Picciotto, Z. Jiang, S.H. Simon, L.N. Pfeiffer, K.W. West, A.M. Sergent, Phys. Rev. Lett. **98**, 206804 (2007)
52. W. Pan, A. Serafin, J.S. Xia, L. Yin, N.S. Sullivan, K.W. Baldwin, K.W. West, L.N. Pfeiffer, D.C. Tsui, Phys. Rev. B. **89**, 241302 (2014)
53. N. Deng, A. Kumar, M.J. Manfra, L.N. Pfeiffer, K.W. West, G.A. Csáthy, Phys. Rev. Lett. **108**, 086803 (2012)
54. N. Deng, J.D. Watson, L.P. Rokhinson, M.J. Manfra, G.A. Csáthy, Phys. Rev. B **86**, 201301 (2012)
55. D.T. Son, Phys. Rev. X **5**, 031027 (2015)
56. M.Z. Hasan, C.L. Kane, Rev. Mod. Phys. **82**, 3045 (2010)
57. X.L. Qi, S.C. Zhang, Rev. Mod. Phys. **83**, 1057 (2011)
58. L.D. Landau, E.M. Lifshitz, *Statistical Physics*, 3rd edn. Part 1. Landau and Lifshitz Course of Theoretical Physics, vol. 5 (Elsevier, Oxford, 1980)

Chapter 3
Low Temperature Measurement Techniques

I have described the interest surrounding the $\nu = 5/2$ FQHS as well as the nematic phase in the third and higher Landau levels. To observe features of these states at all, we need to cool them to milliKelvin temperatures, in order to remove thermal excitations and access the ground state. In order to measure the most sensitive electronic ground states, high quality samples are needed, and they must be studied at temperatures as low as a few milliKelvin. In this section, I will describe the operation of the dilution refrigerator, with which we may obtain milliKelvin temperatures. I will also briefly review some low noise techniques for low temperature measurements, namely, lock-in amplifiers and circuits.

3.1 Dilution Refrigeration

The dilution refrigerator has become a standard instrument for cooling semiconductor materials to milliKelvin temperatures. The basic schematic is depicted in Fig. 3.1. The key component of the dilution refrigerator is the mixture of helium-3 and helium-4 isotopes, but contains many ingredients to ensure the sample is as cold as possible [1–4].

The system is first cooled to 4 K through thermal contact with liquid ^{4}He bath that surrounds the system. The dilution unit will then cool to approximately 1.5 K through the attachment of a ^{4}He cryostat or "1K Pot." This is a small bath of ^{4}He which is pumped on by a rotary pump, reducing the pressure over the bath to around 5 mbar. Attaining this vapor pressure corresponds to decreasing the temperature to around 1.5 K. The 1K pot is constantly replenished by a thin tube connection to the ^{4}He bath. Once cooled to 1.5 K, the ^{3}He/^{4}He mixture is released into the dilution unit, where it condenses and collects in the mixing chamber. When the mixture has been fully condensed into the cold system, it is pumped upon as well and circulated through the system, which cools it even further.

© Springer Nature Switzerland AG 2019
K. A. Schreiber, *Ground States of the Two-Dimensional Electron System at
Half-Filling under Hydrostatic Pressure*, Springer Theses,
https://doi.org/10.1007/978-3-030-26322-5_3

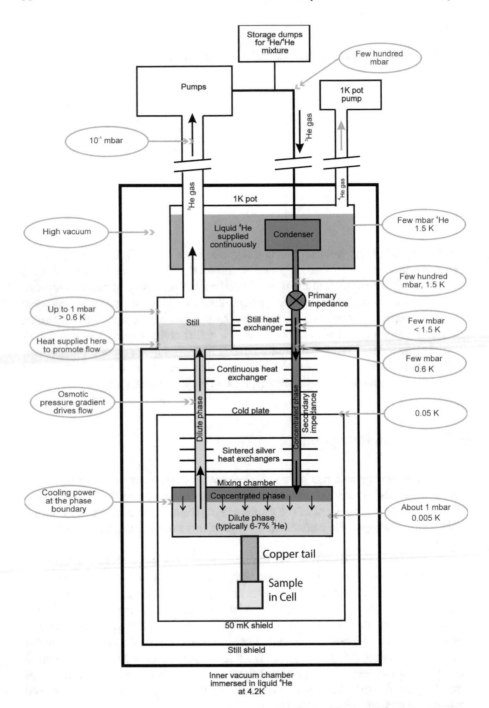

Fig. 3.1 A schematic of the dilution refrigerator, adapted from Ref. [4]. The key component is the mixing chamber, where cooling power is provided by the movement of concentrated ^{3}He (dark blue) across the phase separation boundary into the ^{3}He dilute phase (light blue). The sample in an experimental (yellow) is in thermal contact with the mixing chamber via a copper tail (tan). Image credit to Oxford Instruments NanoScience

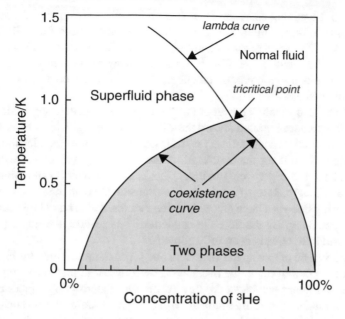

Fig. 3.2 The concentration–temperature phase diagram of ^{3}He/^{4}He mixture, showing the region of the phase separation. Plot from Ref. [4]. Image credit to Oxford Instruments NanoScience

When the mixture is cold enough, it undergoes a phase separation into a ^{3}He rich and a ^{3}He dilute phase. This temperature depends on the concentration of ^{3}He in the ^{4}He, as shown in Fig. 3.2 [1, 4]. The cooling power of the refrigerator is driven by the process of ^{3}He passing from the concentrated to the dilute phase, allowing milliKelvin temperature to be attained. This process shares some thermodynamic similarities with the process of evaporation, but there are some key differences. ^{4}He and ^{3}He are fundamentally quantum liquids, so at zero temperature, ^{3}He still has a *finite* solubility in ^{4}He, around 6.6%. As such, ^{3}He will continue to diffuse into a very dilute mixture of ^{3}He in ^{4}He even at an extremely low temperature. Therefore, this process continues to absorb a latent heat even down to low temperatures. In contrast, the evaporation of a gas into vacuum becomes suppressed at low enough temperature, limiting its ability to cool. The ultimate effect of this argument is that while the cooling power from the evaporation of a gas goes like $\dot{Q}_{\text{evap}} \propto e^{-1/T}$, the cooling power of the passing of ^{3}He across the ^{3}He /^{4}He phase boundary goes like $\dot{Q}_{\text{dilution}} \propto T^2$. That is the key of the dilution refrigerator: the superior ability of the ^{3}He/^{4}He dilution process to cool at very low temperatures.

To continuously run the fridge, ^{3}He must be continuously pulled from the ^{3}He rich phase to the ^{3}He dilute phase. To accomplish this, a small line connects the dilute phase to a chamber called the still. The still contains primarily ^{3}He. It is far removed and thermally isolated from the mixing chamber, and is heated to nearly 1 K. The vapor pressure in the still is therefore much higher than that of the ^{3}He in the mixing chamber. This results in a large osmotic pressure difference between the still and the mixing chamber. Pumping on the still with a powerful pump such as a Roots or a turbo pump, the ^{3}He readily evaporates from the still. Thus ^{3}He is pulled away from the mixing chamber, so more ^{3}He is pulled across the phase boundary to the dilute phase. The ^{3}He boiling away in the still is then collected and re-condenses into the mixing chamber, flowing into the concentrated phase to continue the process.

To maximize the cooling power of this cycle, heat exchangers are needed as well. As the dilute mixture is pulled toward the still, as seen in Fig. 3.1, it passes through a long tube: the heat exchanger. It serves to absorb the heat of the condensing ^{3}He that is returning to the mixing chamber, cooling the returning gas much more effectively than if the heat exchanger were not there. The heat exchanger is often highly coiled to increase its length. Also, to maximize heat transfer, the cold tube is in contact with the condensing line via a large amount of silver sinter, which has a huge surface area to facilitate cooling. By the time the condensed ^{3}He reaches the mixing chamber, ideally it will have already been fully cooled.

The importance of having the chambers and pipes of the refrigerator highly evacuated prior to cooling cannot be overstated. Residual air in the narrow pipes within the fridge will freeze and block the flow of mixture, interrupting a measurement at best and damaging the refrigerator at worst. High vacuum pumps such as turbo pumps are therefore needed to pump out the condenser, still, and all lines that connect to the refrigerator to pressures of 10^{-5} mbar or lower before beginning. Furthermore, traps at liquid nitrogen and liquid helium temperatures are used in sequence to freeze out any air that may leak into the lines over the course of a measurement, and keep it from blocking the flow of the mixture. These ingredients are necessary to a dilution refrigerator, though more components generally play a role in its function, and yet more sophisticated features can be added to play roles in reducing the amount of helium usage.

To attain high magnetic fields of up to 10 T in our refrigerator, a superconducting magnet is needed. The magnet sits in the bottom of the dewar that houses the refrigerator, so is maintained at 4 K at all times during a measurement. It can hold currents of up to 100 A without dissipation. To prevent a quench, which is a condition in which the magnet quickly goes from superconducting to normal and therefore dissipates a huge amount of energy, the field must be ramped slowly and carefully, and the fridge dewar must always be filled with helium.

It is important to know the temperature of the sample for an accurate measurement, so thermometers play an important role. The thermometers we use at milliKelvin temperature on the mixing chamber plate are custom-made carbon resistor thermometers, described in detail in Ref. [5]. A carbon resistor is filed down to a narrow slice, and then embedded in a copper housing, which screws into the mixing chamber. The resistance of this thermometer increases by tens or hundreds of kΩ at the coldest temperatures of a few mK, making it a highly sensitive thermometer.

3.2 Low Noise Electronics

When measuring a sample in a dilution refrigerator, certain considerations must be made, since the sample is at such low temperature [1, 3]. First and foremost, to avoid heating the sample and therefore destroying the fragile ground states that arise, a very low excitation current must be used. Generally, we use $I_{exc} \approx 1 - 10\,nA$. If the sample resistance is on the order of kΩ away from a quantum Hall state, this corresponds to a power dissipation of $P \approx 1 - 10\,fW$, which does not cause the sample to self-heat, and is comfortably below the cooling power of our dilution refrigerator [1]. To measure the small voltage that arises on the sample, on the order of μV, we need sensitive equipment. The workhorse of this measurement is the lock-in amplifier. The lock-in "locks in" on the specified measurement frequency only, allowing for a very low noise measurement. The SRS 830 is typically our lock-in of choice [6]. A low noise pre-amplifier also benefits the measurement by increasing the signal-to-noise ratio. To make the measurement quasi-DC, we choose a low frequency that is not a multiple of the wall frequency (60 Hz), to avoid the noise that comes with this frequency—generally 11 or 13 Hz. Note that the lock-in is a voltage source, but typically in our measurements, it acts as a current source. Adding a large resistor with resistance R on the order of MΩ directly after the output of the lock-in means that all other resistances in the circuit are orders of magnitude lower than this resistor—even Hall resistance is only on the order of tens of kΩ at most. Therefore, the dominant resistance is that of this large resistor, so the current in the circuit is well-approximated by $I \approx V/R$, where V is the voltage sourced by the lock-in. The resistance drop across the sample is then easily attained by $V_{xx} = I R_{xx}$.

A few further notes about the circuit are worth mentioning. For good thermalization of electrons, the wires inside the fridge must be extremely well heat sunk [3]. To this end, the wires are wrapped around copper posts at several stages in the refrigerator in order to cool the electrons en route to the sample. Additionally, ground loops can be a major problem in the measurement. If two points in the circuit are supposed to be ground, but in fact sit at slightly different voltages—indeed, even on the order of μV—the measured result will of course be off. It is thus necessary to ensure the instruments, sample, refrigerator—everything that might be part of the circuit—are at the same ground [3].

A final technique for attaining high quality sample measurements in GaAs at low temperature is that of illumination. Low temperature illumination plays a major role in preparing a high quality sample state. After the sample is placed in the dilution refrigerator, it is warmed to around 10 K before it is cooled to milliKelvin temperatures and measured. Then, light from a red LED illuminates the sample for around 10 min. The sample is then cooled to milliKelvin temperatures for measurement. This technique has been demonstrated to greatly improve the homogeneity of the sample state, allowing for more robust quantum Hall phases to be measured. We employed this technique for each sample that we have measured in this thesis.

3.3   Conclusion

The dilution refrigerator is the central instrument for fractional quantum Hall effect measurements. The samples are cooled to milliKelvin temperatures, thanks to the phase separation process of ^{3}He/^{4}He mixture. Such low temperatures, combined with low noise measurement techniques, allow electron ground states to be observed clearly. All of these low temperature experimental techniques are mandatory for the careful observation of the $\nu = 5/2$ fractional quantum Hall state.

References

1. F. Pobell, *Matter and Methods at Low Temperatures*, 2nd edn. (Springer, Berlin, 1996)
2. D.S. Betts, *An Introduction to Millikelvin Technology* (Cambridge University Press, Cambridge, 1989)
3. J.W. Ekin, *Experimental Techniques for Low Temperature Measurements* (Oxford University Press, Oxford, 2006)
4. N.H. Balshaw, *Practical Cryogenics* (Oxford Instruments Superconductivity Limited, Oxon, 2001)
5. N. Samkharadze, A. Kumar, G.A. Csáthy. J. Low Temp. Phys. **160**, 246 (2010)
6. Model SR 830 DSP Lock-In Amplifier. Stanford Research Systems, Inc. Manual revision 2.5 (2011)

Chapter 4
The Quantum Hall Effect and Hydrostatic Pressure

From the first half of the twentieth century on it has been appreciated that the application of hydrostatic pressure to material systems was an interesting tool for the study of their properties [1–3]. In recent condensed matter experiments, it notably has been used for the tuning of the critical temperature of conventional and of high temperature superconductors [4–7], for driving metal-insulator transitions in various materials [8, 9], and in particular has been used in GaAs heterostructures and quantum wells for the study of the quantum Hall effect [10–14]. Here I review the effect of pressure on GaAs, as well as previous experiments that have been performed in the quantum Hall regime.

4.1 Gallium Arsenide Under Pressure

The general effect of pressure on a crystalline system is to shrink the lattice constant. This has a profound effect on the physics: the Bloch wavefunction is tuned as the lattice constant changes, and so the entire band structure changes. This has most direct impact on the effective mass, effective g-factor, effective dielectric constant, and carrier density. As a result, the quantum Hall regime in GaAs can be studied with the variation of these different parameters.

The dependence of the effective mass and g-factor on the bandgap, valence band spin-orbit splitting, and interband matrix elements in III-V compounds was calculated [15] using $\vec{k} \cdot \vec{p}$ theory. This work formed an important theoretical basis for predicting how these quantities might change with changing band structure. Later, pressure experiments were completed on GaAs, verifying these results, and providing experimental fits for the change in effective mass and g-factor with pressure [10–13].

© Springer Nature Switzerland AG 2019
K. A. Schreiber, *Ground States of the Two-Dimensional Electron System at Half-Filling under Hydrostatic Pressure*, Springer Theses,
https://doi.org/10.1007/978-3-030-26322-5_4

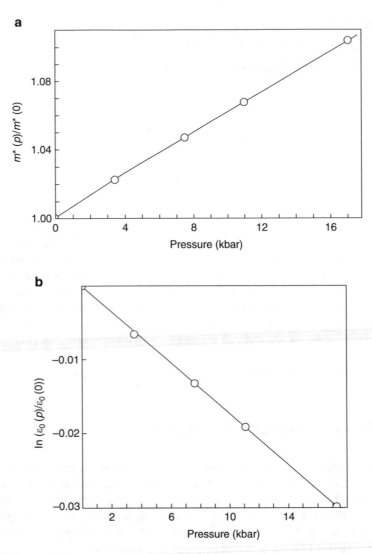

Fig. 4.1 The pressure dependence of two important parameters that change with pressure. (**a**) The effective mass increases with pressure from the ambient pressure effective mass in GaAs. (**b**) The dielectric constant decreases with pressure, from its ambient value in GaAs. Here is plotted $\ln \epsilon(P)/\epsilon(0)$. From Ref. [11], Z. Wasilewski and R.A. Stradling, "Magneto-optical studies of n-GaAs under high hydrostatic pressure." Semicond. Sci. Technol., 1, 264 (1986). ©IOP Publishing. Reproduced with permission. All rights reserved

Reference [11] found through cyclotron resonance measurements that one could fit the effective mass to linear order: $m^*(P)/m^*(0) = 1 + 6.15 \times 10^{-3} P$ where P is measured in kilobar, and $m^*(0)$ is the effective mass used by the authors at zero pressure, equal to $0.0665 m_e$ (Fig. 4.1a). They also derived an experimental fit for the dielectric constant, concluding $d\ln\epsilon(P)/dP = -1.73 \times 10^{-3}$ kbar^{-1} (Fig. 4.1b). Reference [12] calculated an equation for the variation of the g-factor with pressure: $g = -0.43 + 0.0205P$ again where P is in kbar.

It was also recognized early on that the carrier density sharply decreased with pressure [10–14, 18]. It was attributed to the deepening of the donor levels in the bandgap and the relative movement of band minima in GaAs and AlGaAs layers [10, 14, 18].

The rate of density decrease with pressure varies depending on the specific structure of the heterostructure or quantum well and the dopant layers, and some of the density can be recovered after pressurizing upon illuminating the sample [10–14]. In general, however, the density decrease is one of the most significant changes in the pressurized samples, and, as we shall argue, likely plays the dominant role in determining the energy scales in our own experiment.

As the pressure increases, the mobility decreases as well. Figure 4.2 displays our experimental data showing how the density and mobility decrease with pressure. We determine that the density decrease is 2.17×10^{10} cm^{-2} per kilobar.

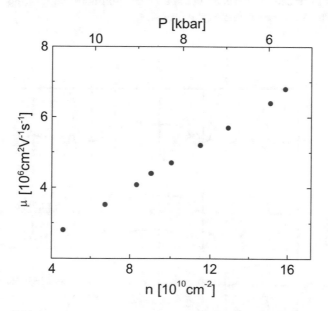

Fig. 4.2 As pressure increases, electron density in the sample decreases linearly, and mobility decreases as well. Pressure is plotted on the top axis, while density is plotted on the lower axis. This sample was studied in Ref. [16, 17]. Data from the supplement of Ref. [16]. Reprinted figure with permission from K.A. Schreiber et al., Phys. Rev. B 96, 041107 (2017). Copyright 2017 by the American Physical Society

4.2   Previous Experiments of the Fractional Quantum Hall Effect Under Pressure

In the 1980s and 1990s, several experiments examined quantum Hall states in the lowest Landau level under hydrostatic pressure. Because of the capability of hydrostatic pressure to tune the g-factor, its utility in driving spin transitions was primarily explored. The degree to which a fractional quantum Hall state is spin polarized is determined by the magnitude of the Zeeman energy. In the limit of zero Zeeman energy—that is, first imagining the g-factor to be zero—the energy spectrum consists of spin-degenerate Landau levels and composite fermion lambda levels. As the Zeeman energy increases, either by tuning the g-factor or increasing the total magnetic field at a fixed filling factor by increasing an in-plane magnetic field, the spin degeneracy is lifted: the spin-up states decrease in energy and the spin-down states increase in energy. Eventually, there is a crossing of energy levels. At this point the gap closes, and we observe a spin transition. Hence at the point of spin transition, we will observe no apparent quantum Hall minimum. Moving away from the spin transition, we then see the FQH minimum reappearing, as the gap re-opens. This sequence is depicted in Fig. 4.3.

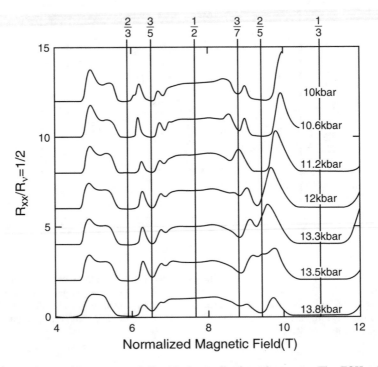

Fig. 4.3 A spin transition at $\nu = 2/5$ with the application of pressure. The FQH minimum disappears, then reappears as the g-factor is increased. This is moving through a spin transition with tuning of the Zeeman energy. Plot from Ref. [24]. Reprinted figure with permission from W. Kang et al., Phys. Rev. B 56, 12776 (1997). Copyright 1997 by the American Physical Society

It is not always immediately obvious if a FQHS is spin polarized or not [19, 20]. At fractional quantum Hall values, the Zeeman energy in GaAs is of the same order as the energy spacing between composite fermion levels, so a given state may be spin polarized or unpolarized, depending on the structure of the composite fermion levels. In particular a state may undergo a spin transition from a polarized to unpolarized state or vice versa.

One of the earliest experiments utilizing hydrostatic pressure to demonstrate that a fractional quantum Hall state was unpolarized was a study at $\nu = 4/3$ [21, 22]. A spin transition in this state had already been observed under tilted field experiments [23], and the pressure results confirmed this experiment. 5/3 and 7/5 were also seen to be weakly enhanced by pressure [22]. Further experiments led to observations of spin transition at $\nu = 2/5$ [24], as well as near 4/7 and 4/9, with some surprising hysteretic behavior near these fractions [25] (Fig. 4.3). Therefore the g-factor tuning proved to be powerful in driving spin transitions in the FQH regime.

Other unusual spin effects arise when one tunes the g-factor down towards zero, which occurs at $P = 17$ kbar. One such interesting effect is the presence of charged spin texture excitations, also referred to as Skyrmions, at filling factor 1 and at filling factors in the Jain hierarchies such that $\nu^* = 1$ [26–28]. The Skyrmions are objects experimentally observed to have a typical radius of several magnetic lengths and spin on the order of 10. They arise in an energy regime where it happens that a higher Zeeman energy is a favored state for the benefit of lowering the exchange energy. Experimentally, hydrostatic pressure experiments confirmed the presence of high-spin quasiparticles at $\nu = 1$ [26]. Tuning the Zeeman energy through zero and noting that the slope of the gap as a function of total B-field could be given by $\partial \Delta / \partial B_{tot} = s\mu|g|$, where s is the total spin of the excitations, they were able to confirm a drastically increasing s as g-factor went through zero, up to $s = 33$. A similar experiment suggested the presence of Skyrmions at $\nu = 1/3$ as well [27].

4.3 Pressure Clamp Cell

To pressurize the sample in a way that is suitable for a quantum Hall measurement, several factors must come into play. The pressure must be hydrostatically applied, even if the medium which applies the pressure freezes, to avoid inducing unintentional anisotropy. Second, to maximize the pressure applied, the sample and wires must fit into as small an area as feasible so that one may apply a relatively small force. All of these factors are considered in our experiment.

The cell we use is a clamp cell from Almax easyLab, model Pcell 30 [29]. Clamp cells are suitable for attaining relatively low pressures in condensed matter systems, less than 100 kbar. Pressure is applied by depressing a piston into a cylinder, compressing the sample within, which is immersed in a fluid or compacted powder. The cell is then clamped, typically by tightening a nut on the cell, to hold the pressure within. The cells may be made of beryllium copper, tungsten carbide, or a proprietary alloy of one of these, so that the cell may withstand the desired pressure.

Our own pressure cell is depicted in Figs. 4.4 and 4.5. It is made of beryllium copper and proprietary alloys and can withstand up to 30 kbar. It holds the sample, which is 2 mm × 2 mm, as well as an LED, which is used for standard low-temperature illumination techniques to improve the homogeneity of the sample. We also include two manometers to let us determine the pressure. At room temperature, we use the resistance of a manganin wire, which is sensitive to pressure, as our indicator, and at low temperature, we use the superconducting transition of tin, which is sensitive to pressure, as our indicator. The low temperature pressure we measure is consistently about 5 kbar lower than that which we measure at room temperature, due to the freezing of our hydrostatic pressure-transmitting fluid. The fluid that we use is an equal mixture of pentane and isopentane, which, over our pressure range of interest, freezes isotropically at cryogenic temperatures [30]. (Note that pure pentane or isopentane is found to solidify at *room* temperature at $P = 18$ kbar [30]).

All of these electrical components fit into a Teflon tube about 3 mm in diameter, which is filled with the pressure-transmitting fluid. The wires are guided out

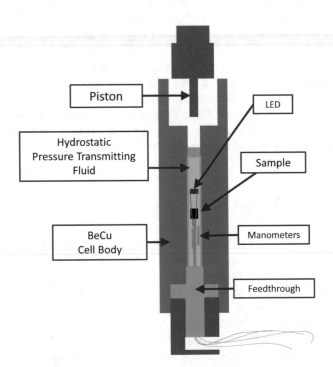

Fig. 4.4 Schematic of the pressure cell and setup of the sample, manometers, and LED within the cell. The sample, manometers, and LED are mounted on the feedthrough and fit into the Teflon feedthrough cover with the hydrostatic pressure-transmitting fluid. The feedthrough is inserted into the cell, and pressure is applied by displacing the piston

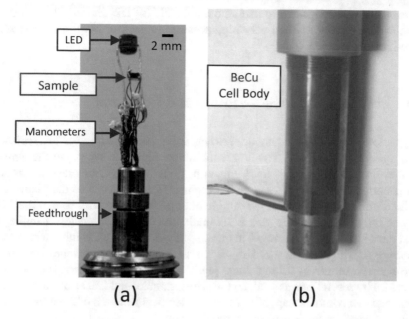

Fig. 4.5 (**a**) A photograph of the sample, manometers, and LED, mounted to the feedthrough that is inserted into a Teflon tube and then into the pressure cell. (**b**) The pressure cell itself mounted to a tail in preparation for insertion into the dilution refrigerator

through a feedthrough. The Teflon tube slides into the pressure cell, and locking nuts are screwed into top and bottom. Then, high pressure is applied using a hydrostatic pressure ram. A tungsten carbide piston compresses the Teflon tube and the components within. As the piston is depressed, the lower locking nut is screwed in a few turns at a time, so that when the hydrostatic pressure ram releases its pressure, the cell remains under that attained pressure.

4.3.1   Diamond Anvil Cells

Diamond anvil cells are used for pressures much higher than used in our own experiments, but because of their widespread use and great utility for achieving very high pressure, I will mention them here. The best diamond anvil cells may withstand hundreds of gigapascals of pressure (note that 1 GPa = 10 kbar!) making them suitable for experiments which drive structural transitions, such as a recent experiment claiming the observation of metallic hydrogen [31], or experiments which attempt to maximize the critical pressure in high T_c superconductors [4–7]. A high quality diamond is used as the cap in the diamond anvil cell. Importantly, for most diamond anvil cells, experimental signatures must be obtained contactlessly, as wire feedthroughs reduce the highest safely attainable pressure in such cells. Optical experiments are well suited to diamond anvil cells for this reason. Thus, diamond anvil cells are suitable when very high pressure is needed, but limit the types of experiments possible.

4.4 Preparing for Pressurization and Cooldown

In practice, several issues may arise when preparing a sample for pressurization in our pressure clamp cell. The first is the mounting of the sample to the wires of the feedthrough. The sample must remain parallel to the ground when the pressure cell is upright, so that the magnetic field remains perpendicular to the plane of the 2DES. Furthermore the top of the sample must be facing upward, so that the 2DES may be fully illuminated by the LED. Finally, the wires, manometers, sample, and LED must be compactly arranged in order to fit within the Teflon tube, with enough clearance to withstand the shrinking of the Teflon tube under pressure. The Teflon tube diameter shrinks by 0.1 mm under pressurization of several kbar, and the length decreases by several mm as well, so the wires, sample, and LED should not extend more than 6 mm into the cap [32]. However, electrical isolation of the wires must be maintained despite these strict space requirements.

Satisfying these demands is not trivial, and successfully soldering a sample in the correct arrangement may take days of patient manipulation under the microscope. A few general tips are recommended. It is essential that the wires and solder remain clean and unoxidized. As low a temperature for soldering as possible should be used, and the soldering iron tip should be frequently cleaned with a clean glass slide. Unnecessary bending of the wires on the feedthrough is strongly discouraged, as they are quite fragile and will break. If too many wires break, the feedthrough will be unusable. To this end it is advisable to make sure all wires that connect to the sample should be of the same length, and in the positions you want them to be, before beginning. I provide a detailed procedure for successfully mounting the sample to the feedthrough here.

4.4.1 Mounting the Sample to Pressure Cell Feedthrough

The following steps led me to a successful and relatively efficient mounting of the sample. Completing the steps in this order will make it less likely you need to go back and adjust the position of the components in your feedthrough, which increases the likelihood your wires will break. Work under the microscope to maximize your precision. It is recommended to practice first with wires of the same gauge arranged in the same configuration as a feedthrough. For example, an old feedthrough that did not survive pressurization is suitable.

First, ensure that the prewired manometers are positioned near the base of the feedthrough and are tucked in sufficiently to avoid the Teflon cap. If you choose to change the tin manometer that is provided by the company with another pressure-sensitive superconducting material, do this now, being careful not to excessively bend any wires. If possible, wrap the exposed parts of the metal wire with tiny pieces of Teflon tape. Complete a resistance check at the other end of the wires to ensure all connections are good. I recommend soldering a connector at this end of the wires from the beginning, so you can complete easier electrical checks as you go. The two pairs of blue and red wires with red beads are for a four terminal measurement of the manganin, and the two pairs with yellow beads are for the superconducting manometer. The resistance of the manganin should be around 20–25 Ω. Be gentle with the manometers, and once they are out of your way and you have ensured they are well-soldered, do not move them again.

Next, choose the wires you want to use for your sample and the LED. This takes a bit of strategizing. The four wires that will be used for the sample contacts absolutely must be at the same height, or else the sample will be tilted. Fortunately, the provided feedthroughs have at least two pairs of the empty wires cut to the same length, so your task is easier, as long as you do not later break a wire. Gently untwist the pairs a few turns. Arrange them so that the wire tips form a square about the size of your sample, well-centered along the feedthrough's axis. It is useful to have the Teflon cap handy, so that you can often check to make sure your wires will clear the cap when you put it on. If you have a fresh new feedthrough, the insulation is already removed from the wire tips, and pre-tinned, likely with lead solder. Pre-tin these tips now yourself with indium solder at 360 °F, as indium solder is needed for your sample's contacts. Use very fresh solder, and do not add flux. Now carefully bend the wire tips inward, just slightly, so that they will overlap with the contacts of your sample.

When these wires are arranged suitably, choose the next two wires for the LED and mount the LED before mounting the sample. I have found greatest success when the sample is the last thing mounted, so that there is a smaller chance of knocking the sample out of position, tilting it, or disconnecting a sample contact. The wires you choose for the LED will need to be at the outermost edges of the feedthrough, as in the Fig. 4.5. Gently untwist them and position them as such.

The LED should be relatively small and flat, like the type we have used in the figure. Ensure that its diameter is small enough to fit in the Teflon cap, with clearance to account for the shrinkage of the cap. You may need to trim the corners of the LED carefully using a razor blade. The LED will go above the sample, and its face will point downward to shine on the sample. With this in mind, gently bend the legs of the LED entirely forward. Now, keeping in mind that electrical components need to be far away from the end of the Teflon tube to account of the Teflon shrinkage, trim the LED legs to the length you want. Ideally, the insulated part of the wires you chose for the LED will be at the same height as where your sample's contacts will be, so that lateral movement of the wires in the feedthrough is less likely to result in shorting your sample to the LED, but this may not always be possible, depending on the length of the wires. If needed you can consider extending the feedthrough wires by soldering short lengths of copper wire to them, or wrapping thin sheets or tubes of kapton around solder joints and the LED legs. However, I found that simply wrapping a small piece of Teflon tape around the sample perimeter after it was soldered was sufficient to ensure the sample does not short to the LED.

When the LED wires are prepared and the LED legs are cut to the correct length, solder on the LED. Though indium solder is always used for the sample, lead solder can be used to attach the LED. When it is attached, and you have done an electrical check, and ensured that the LED is cleared by the Teflon cap, gently push the LED out of the way at a slight angle, without bending the wires too much. It is time to attach the sample.

This is the most difficult part, so have a fresh mind going into it. Use a soldering iron temperature of 360 °F. The sample will ultimately be suspended by the four wires you have chosen, and again, must remain perpendicular to the magnetic field that will be applied. To put the sample in position, use a thin wooden stick with a pointy tip. Carve a flat spot on the tip on which the sample can rest. Use a third hand to hold this wooden stick so the flat spot is parallel to the ground. Use a tiny amount of rubber cement, and attach the sample to this flat spot, without getting rubber cement on the sample face. Let the rubber cement dry for about 15 min.

When the sample is secure enough not to fall off, and is in position parallel to the table, carefully move the stick so the sample is between the four wires you had previously prepared. With tweezers, make sure the four wire tips are overlapping your sample contacts. Now, with a clean soldering iron tip and tweezers, press each wire tip to your contact. Try to apply the heat for as short a time as possible, to avoid oxidizing your contacts or heating up the sample so much that contacts you have already soldered melt.

When the contacts are soldered, do an electrical check. The sample resistance should be on the order of a few kΩ. If everything seems to work out, use tweezers to *extremely* gently nudge the sample repeatedly until it detaches from rubber cement. If your soldering joint was good, this small perturbation will not detach your sample. When the sample is free, remove the wooden stick from the area. Now carefully check that the sample is still parallel to the ground. It should be very close if you arranged the wires well before you began. Very carefully nudge the sample and wires if it needs adjustment. When you are satisfied, complete another electrical check of the sample.

When the sample is soldered and arranged, cut a small piece of Teflon tape a few millimeters wide. Using tweezers, lightly wrap the perimeter of the sample with it. Make sure the contacts are covered from the side, but do not cover the sample top with it. Now move the LED back up so that it is directly over the sample. Do another electrical check of all components, making sure that no wires short to any other wires.

Now make sure the Teflon cap slides on. Do this very carefully. If everything has been arranged well, no wires will be touched. In the worst case, the Teflon cap will catch one of your sample wires and bend it, completely tilting your sample or even breaking the wire contact. If this happens you will have to solder it again. If the Teflon cap slides on without touching anything, do yet one more electrical check. If everything checks out, well done! Leave the cap on for protection and store the feedthrough safely until you are ready to insert the pressure transmitting fluid and pressurize the sample. The fewer times you have to take the Teflon cap on and off, the less chance there will be that the sample will be disturbed.

4.5 Monitoring the Effect of Pressure

To determine the pressure inside the cell, two types of manometers, or pressure gauges, are provided by Almax easyLab. Properties of the sample and LED also change with pressure and can be monitored for a sense of the effect of pressure.

4.5.1 Room Temperature Pressure Monitoring

The pressure at room temperature must be monitored as it is applied in order to attain the target pressure. To this end, a small manganin wire is provided, mounted in the feedthrough by the company. Its resistance changes with pressure at room temperature, permitting us to monitor the pressure as it is applied. Manganin's response to pressure is given by [32]

$$P = 403.23\left[\frac{R(P)}{R(0)} - 1\right] \tag{4.1}$$

where $R(0)$ is the zero pressure resistance and P is measured in kbar. Decent sensitivity in the measurement of this resistance is needed: the provided manganin has a resistance of about 25 Ω, so the resistance only increases by an ohm or less over a typical pressure range of 10–15 kbar. A bad solder joint or contact to the multimeter can therefore result in an inaccurate reading.

It is desirable, due to the insensitivity of the provided manganin, and as a backup in the case that wires to the manganin break over the course of an extended pressure campaign, to have another indicator of the pressure at room temperature. For our GaAs samples, it happens that the sample's own resistance is extremely sensitive to pressure. Monitoring the sample's resistance in fact permits us to take small pressure steps, attaining target pressures with a higher degree of accuracy than monitoring the manganin alone. Figure 4.6 shows the two-terminal and four-terminal resistance change of one of our GaAs samples, referred to later as sample 2, always measured with the same respective sets of contacts. The resistance change is likely tied to the decrease in carrier density and increased trapping potential of ionized donors [10]. Due to the differences in the curves of the two-terminal and four-terminal resistances, it appears ohmic contacts have their own sensitivity to pressure, which can be exploited as a pressure gauge as well. The precise mathematical fit of the resistance change with pressure may be sample dependent, relying on the sample size and growth parameters such as doping setback and quantum well width. A linear extrapolation from the previous few pressure-sample resistance points, however, tends to yield a very accurate prediction of the pressure of the next point.

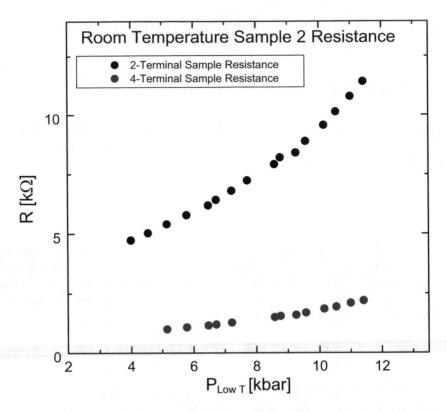

Fig. 4.6 The dependence of a GaAs sample's four-terminal and two-terminal resistance at room temperature on the pressure attained at low temperature, about 5 kbar lower than that at room temperature. Note that the ohmic contacts have a response to pressure, as evidenced by the difference of the four-terminal and two-terminal curves. This sensitive dependence on pressure makes the sample's room temperature a good secondary pressure gauge

The displacement of the piston within the cell gives us a measure of how much the Teflon tube encapsulating the feedthrough has been compressed, though this translates to only an approximate estimate of the pressure. Figure 4.7a shows a plot of the pressure dependence on piston displacement, d. It is mandatory to monitor the piston displacement during the pressurization process by estimating how many times the locking nut is turned, as the pressure cell can be damaged if the locking nut is overtightened. After the target pressure is attained, the piston displacement must be measured with a caliper by measuring the remaining height of the locking nut protruding from the bottom of the cell. Displacement of the piston d is obtained by subtracting the height of the locking nut $Z(P)$ (Fig. 4.7c) from the height of the locking nut at zero pressure, $Z(0)$ (Fig. 4.7b). The more the locking nut has been screwed in, the farther the piston has been displaced, and the higher the pressure. However, a pressurization followed by a depressurization step leads to some hysteresis. Once compressed, the Teflon cap does not perfectly return to its

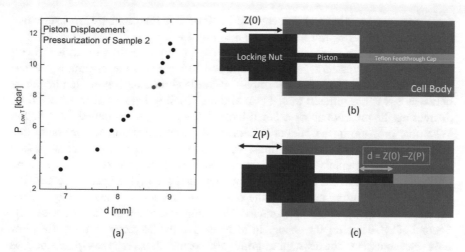

Fig. 4.7 (**a**) The low temperature pressure dependence of the piston displacement. As the locking nut is screwed in, and as the piston therefore compresses the Teflon cover of the feedthrough within, the pressure inside the feedthrough increases. (**b**) The locking nut height at zero pressure, $Z(0)$, measured with calipers. (**c**) Measuring the piston displacement d by obtaining $Z(P)$ after each pressurization

original size. Hence, the piston needs to be displaced slightly further to reattain the higher pressure.

At room temperature, the opening voltage of the LED is hardly sensitive to pressure at all. While it is necessary to monitor the LED's opening voltage at room temperature after a pressurization to ensure it still works and no wires have broken, one is likely to see little change until the LED is cooled.

4.5.2   Low Temperature Pressure Monitoring

When the sample is cooled to a few Kelvin and below, the pressure is lower than that at room temperature. The pentane/isopentane pressure transmitting fluid freezes hydrostatically [32], resulting in a pressure about 5.7 kbar lower than that measured at room temperature. This pressure may be monitored by measuring the superconducting transition temperature T_c of a metal included in the pressure cell. Almax easyLab provides a short tin wire for this purpose, which is set up for a resistive measurement. Tin's T_c decreases from 3.73 K at zero kbar to around 2.50 K at 30 kbar [32], although the exact T_c depends on the impurity content of the metal as well as any stray magnetic fields. The equation used to extract the pressure is given by [32]

$$P = 5.041[T_c(0) - T_c(P)]^2 + 17.813[T_c(0) - T_c(P)] \tag{4.2}$$

for P measured in kbar and T measured in Kelvin. A representative measurement of the superconducting transition is given in Fig. 4.8.

The resistance measurement of the superconducting transition of tin is difficult for a handful of reasons. First, in a dilution refrigerator, the apparatus we use that is ideally suited to measure samples at milliKelvin temperature, temperatures between 2–4 K are difficult to stably maintain. Cooling below 4 K is achieved by decreasing the pressure above a small bath of liquid helium-4, called the 1 K pot. When the pressure in the 1K pot is decreased to several millibar, a pressure that is relatively easily attainable by most vacuum pumps, the temperature decreases to about 1.5 K. Maintaining temperatures above 1.5 K but below 4 K is therefore difficult to control in our apparatus, because it entails applying heat to increase the vapor pressure, putting a strain on the pump. For an accurate measurement of the superconducting transition temperature, the temperature should be swept slowly. Second of all, detecting the change in resistance in the tin requires a sensitive, low noise measurement. The resistance of tin in the normal state is already quite small, as it is a metal. The measurement must be able to resolve therefore a resistance change of a few $\mu\Omega$. Our lock-in amplifiers and low-noise circuitry are up to the task, as displayed in Fig. 4.8. Alternative measurements are worth considering, however, to simplify and shorten the measurement.

A plan for a more suitable low temperature manometer is to replace the piece of tin with another superconducting metal, and to measure the change in its inductance, rather than its resistance. Two metals may be suitable for our purposes: zinc and

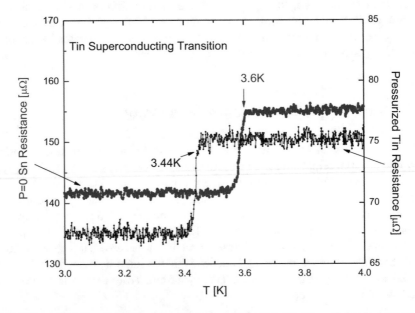

Fig. 4.8 Representative measurements of the superconducting transition of the tin manometers provided in the feedthrough. The tin's resistance is measured by lock-in amplifier as temperature is slowly varied. The red trace is a measurement at zero pressure, while the black trace is a measurement at about 3 kbar

lead. Zinc has a T_c varies from 900 mK to 300 mK from 0 to 20 kbar, which is easily attainable under normal dilution fridge operating conditions [33]. Lead, whose T_c varies between 7 and 5 K from 0 to 40 kbar [33], is also attractive, because it can be placed in a bath of helium-4 and warmed to the desired temperature without need for vacuum pumps at all. The superconducting transition entails not only a large drop in resistivity but also a sudden change in the magnetic susceptibility. This means the transition can be detected through a change in the material's inductance. A small inductor can be made and included in the cell by wrapping a copper wire around a piece of the metal. The inductance can then be measured using lock-in techniques. These alternative inductors may be explored for low temperature pressure determination.

When the sample is at low temperature, we are able to obtain its density through magnetotransport measurements. The magnetic field of a known filling factor with a narrow R_{xx} minimum, such as $v = 11/7$, can be measured, and from that we may accurately extract the density from the definition of filling factor: $v = hn/eB$. As discussed above, the density decreases linearly with the increasing pressure. We observe a decrease of density of $dn/dP = -2.17 \times 10^{10}$ cm^{-2}/kbar. This can be used as an additional gauge of the pressure at low temperature.

The LED's opening voltage is sensitive to pressure at low temperature, but is dependent on the LED. The voltage drop across the LED in our first series

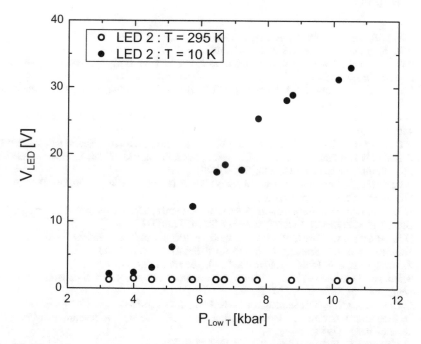

Fig. 4.9 The voltage response of the LED in the measurement of sample 2 with pressure at room temperature at $T = 10$ K. At room temperature, the voltage presented here is the opening voltage, measured using a Fluke digital multimeter. At 10 K, 1 mA was sourced to the LED and the corresponding voltage was measured

of pressurizations saturated at around 20 V at pressures above 10 kbar. In the pressurization of the second sample, the LED's voltage drop grew to surpass 30 V when sourcing 1 mA. The voltage response at room temperature and 10 K in this LED is presented in Fig. 4.9. Above 11 kbar, the voltage drop suddenly approached 70 V even when sourcing 10 μA in this LED. Possibly the LED was damaged at this pressure. An LED is not a reliable gauge at low temperature, due to differences in individual LEDs.

4.6 Conclusion

Hydrostatic pressure is of great use for changing sample band parameters. The use of our hydrostatic pressure clamp cell lets us attain up to 30 kbar, enough to change many relevant parameters that will affect the energy scales experienced by fractional quantum Hall states. We are therefore interested in studying the state $\nu = 5/2$— and indeed, the higher Landau levels, which have not yet been explored—under pressure.

References

1. P.W. Bridgman, Rev. Mod. Phys. **18**, 1 (1946)
2. P.W. Bridgman, Proc. Phys. Soc. **41**, 341 (1928)
3. P.W. Bridgman, General survey of certain results in the field of high-pressure physics. Nobel Lecture, Dec. 11, 1946
4. S.M. Souliou, A. Subedi, Y.T. Song, C.T. Lin, K. Syassen, B. Keimer, M. Le Tacon, Phys. Rev. B **90**, 140501 (2014)
5. X.-J. Chen, V.V. Struzhkin, Y. Yu, A.F. Goncharov, C.T. Lin, H.-K. Mao, R.J. Hemley, Nature **466**, 950 (2010)
6. A.P. Drozdov, M.I. Eremets, I.A. Troyan, V. Ksenofontov, S.I. Shylin, Nature **525**, 73 (2015)
7. F. Hardy, N.J. Hillier, C. Meingast, D. Colson, Y. Li, N. Barisic, G. Yu, X. Zhao, M. Greven, J.S. Schilling, Phys. Rev. Lett. **105**, 167002 (2010)
8. P. Postorino, A. Congeduti, P. Dore, A. Sacchetti, F. Gorelli, L. Ulivi, A. Kumar, D.D. Sarma, Phys. Rev. Lett. **91**, 175501 (2003)
9. M.D. Knudson, M.P. Desjarlais, A. Becker, R.W. Lemke, K.R. Cochrane, M.E. Savage, D.E. Bliss, T.R. Mattsson, R. Redmer, Science **348**, 1455 (2015)
10. D.K. Maude, J.C. Portal, Parallel transport in low-dimensional semiconductor structures. *Semiconductors and Semimetals*, vol. 55 (Elsevier, Amsterdam, 1998)
11. Z. Wasilewski, R.A. Stradling, Semicond. Sci. Technol. **1**, 264 (1986)
12. S. Holmes, D.K. Maude, M.L. Williams, J.J. Harris, J.C. Portal, K.W.J. Barnham, C.T. Foxon, Semicond. Sci. Technol. **9**, 1549 (1994)
13. L. Dmowski, J.C. Portal, Semicond. Sci. Technol. **4**, 211 (1989)
14. G. Gregoris, D. Lavielle, J. Beerens, S. Ben Amor, J.C. Portal, F. Alexandre, Semicond. Sci. Technol. **4**, 317 (1989)
15. C. Hermann, C. Weisbuch, Phys. Rev. B **15**, 823 (1977)
16. K.A. Schreiber, N. Samkharadze, G.C. Gardner, R.R. Biswas, M.J. Manfra, G.A. Csáthy, Phys. Rev. B **96**, 041107 (2017)

17. N. Samkharadze, K.A. Schreiber, G.C. Gardner, M.J. Manfra, E. Fradkin, G.A. Csáthy. Nat. Phys. **12**, 191 (2016)
18. J. Beerens, G. Gregoris, J.C. Portal, J.L. Robert, J.M. Mercy, F. Alexandre, Semicond. Sci. Technol. **3**, 577 (1988)
19. J.K. Jain, *Composite Fermions* (Cambridge University Press, Cambridge, 2007)
20. R.E. Prange, S.M. Girvin, *The Quantum Hall Effect* (Springer, Berlin, 1987)
21. N.G. Morawicz, K.W.J. Barnham, A. Briggs, C.T. Foxon, J.J. Harris, S.P. Najda, J.C. Portal, M.L. Williams, Semicond. Sci. Technol. **8**, 333 (1993)
22. N.G. Morawicz, K.W.J. Barnham, C. Zammit, J.J. Harris, C.T. Foxon, P. Kujawinski, Phys. Rev. B **41**, 12687 (1990)
23. R.G. Clark, S.R. Haynes, A.M. Suckling, J.R. Mallett, P.A. Wright, J.J. Harris, C.T. Foxon. Phys. Rev. Lett. **62**, 1536 (1989)
24. W. Kang, J.B. Young, S.T. Hannahs, E. Palm, K.L. Campman, A.C. Gossard, Phys. Rev. B **56**, 12776 (1997)
25. H. Cho, J.B. Young, W. Kang, K.L. Campman, A.C. Gossard, M. Bichler, W. Wegscheider. Phys. Rev. Lett. **81**, 2522 (1998)
26. D.K. Maude, M. Potemski, J.C. Portal, M. Henini, L. Eaves, G. Hill, M.A. Pate, Phys. Rev. Lett. **77**, 4604 (1996)
27. D.R. Leadley, R.J. Nicholas, D.K. Maude, A.N. Utjuzh, J.C. Portal, J.J. Harris, C.T. Foxon, Phys. Rev. Lett. **79**, 4246 (1997)
28. A. Schmeller, J.P. Eisenstein, L.N. Pfeiffer, K.W. West, Phys. Rev. Lett. **75**, 4290 (1995)
29. easyLab Technologies Ltd., model Pcell 30
30. S.H. Gelles, J. Chem. Phys. **48**, 526 (1968)
31. R.P. Dias, I.F. Silvera, Science **355**, 715 (2017)
32. easyLab Pcell 15/30 User Guide, easyLab technologies Limited (2004)
33. P.E. Seiden, Phys. Rev. **179**, 458 (1969)

Chapter 5
The Fractional Quantum Hall State-to-Nematic Phase Transition Under Hydrostatic Pressure

Having introduced the electronic phases in the 2DES and the utility of hydrostatic pressure to tune these phases, I now turn to our experiment to observe the second Landau level under hydrostatic pressure. Prior published work on the FQHSs in pressurized GaAs did not present fruitful results on the second Landau level. Sample mobility has improved dramatically since those experiments were completed, and we have the ability to include an LED in our cell to prepare the sample state after each pressurization. We therefore took advantage of a unique opportunity to probe the second Landau level in ways not previously explored. We made an unexpected discovery: with increasing pressure, the FQHS at $\nu = 5/2$ weakened and disappeared, and gave way to a nematic phase above a critical pressure. This marks the first observation of a nematic phase at $\nu = 5/2$ that did not arise as a result of a externally applied field that explicitly breaks rotational symmetry, such as an in-plane magnetic field. Furthermore, this phase transition from FQHS to nematic phase is a rather unusual one in that two kinds of order change in the transition. The nematic order, a conventional order well-described by Landau's theory of phase transitions, is acquired with the increasing pressure, and the topological order of the FQHS is lost. We construct a diagram summarizing the phases in pressure–temperature space. We also present evidence that the FQHS-to-nematic transition is a quantum phase transition: a transition at zero temperature. Quantum phase transitions exist throughout condensed matter physics and are of a great deal of interest, as they may shine light on the way different phases influence one another. Our results may provide further insight into the way the paired, topological $\nu = 5/2$ may influence the nematic phase. The chapter is very similar to work published in Refs. [1] and [2].

© Springer Nature Switzerland AG 2019
K. A. Schreiber, *Ground States of the Two-Dimensional Electron System at Half-Filling under Hydrostatic Pressure*, Springer Theses,
https://doi.org/10.1007/978-3-030-26322-5_5

5.1 Observation of the Fractional Quantum Hall State-to-Nematic Transition at $\nu = 5/2$

The first sample that we measured in the pressure cell was from the same wafer of the sample that was studied in Ref. [3] under ambient pressure. Throughout this thesis, I will refer to this sample as sample 1. The sample is a 30 nm quantum well, with ambient density $n = 2.8 \times 10^{11}$ cm^{-2} and mobility $\mu = 15 \times 10^{6}$ cm/Vs. It was cleaved to 2×2 mm^2 so that it can easily fit inside the Teflon lining of the pressure cell. We pressurized the sample and loaded it into our dilution refrigerator, using the techniques described in the previous chapters. We performed Hall measurements using standard low-frequency lock-in amplifier techniques with an excitation of 2 nA. After each pressurization and cooldown, the sample was illuminated at around 10 K for 10–20 min.

In Fig. 5.1, we show the longitudinal resistance measured at three different pressures at about 12 mK along the perpendicular crystallographic directions of the GaAs. R_{xx} is obtained from the current bias applied and the voltage drop measured along the $\langle 1\bar{1}0 \rangle$ crystal direction, whereas R_{yy} is measured along the $\langle 110 \rangle$ direction.

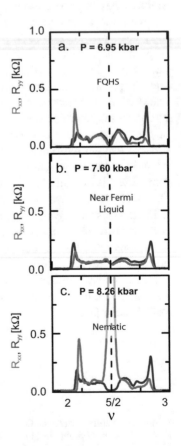

Fig. 5.1 The green traces show R_{xx} and the red traces show R_{yy}, as measured along two mutually perpendicular crystallographic directions of GaAs. R_{xx} is measured along the crystallographic direction $\langle 1\bar{1}0 \rangle$ and R_{yy} is measured along $\langle 110 \rangle$ As the pressure is increased, at $\nu = 5/2$ we observe the following sequence of ground states: an isotropic FQHS (**a**), a nearly isotropic Fermi liquid (**b**), and the nematic phase (**c**).The data is taken at $T \simeq 12$ mK. Plots from Ref. [1]

In Fig. 5.1a, we show the magnetoresistances at $P = 6.95$ kbar. Both R_{xx} and R_{yy} exhibit sharp minima at $\nu = 5/2$, and R_{xx} and R_{yy} measured in the vicinity of $\nu = 5/2$ along the different sample edges are nearly equal. We thus conclude that, as in measurements performed on samples in the ambient, the ground state is an FQHS at $\nu = 5/2$ and $P = 6.95$ kbar. As the pressure is increased to $P = 7.60$ kbar, the longitudinal resistance near $\nu = 5/2$ remains isotropic. However, as seen in Fig. 5.1b, the strong minima in R_{xx} and R_{yy} are no longer present. The finite and isotropic resistance at $\nu = 5/2$ is reminiscent of a compressible isotropic Fermi liquid. This suggests that the ground state at $\nu = 5/2$ approaches an instability.

A further increase in pressure to $P = 8.26$ kbar causes a strong minimum to reappear in R_{yy} at $\nu = 5/2$. As seen in Fig. 5.1c, this minimum in R_{yy} is visibly wider—in that it spans a larger range of filling factor—than that at $P = 6.95$ kbar. The most marked change, however, is in R_{xx}, which exhibits a pronounced peak at $\nu = 5/2$. The anisotropic resistance observed at $\nu = 5/2$, characterized by an extremely large ratio $R_{xx}/R_{yy} = 1, 150$, signals the onset of a ground state which breaks rotational symmetry. The evolution of the magnetoresistance at $\nu = 5/2$ shown in Fig. 5.1 is therefore suggestive of a phase transition from the rotationally invariant $\nu = 5/2$ FQHS, which as we discussed in chapter one is most likely a non-Abelian topological phase [4–6], to an anisotropic phase. We note that in Fig. 5.1 resistance anisotropy develops not only at $\nu = 5/2$, but also at filling factors close to $\nu = 2.2$ and 2.8. However, in contrast to the anisotropy at $\nu = 5/2$, that at $\nu = 2.2$ and 2.8 is not sensitive to the temperature, and it is commonly associated with geometric imperfections of the sample and of the contact placement [7, 8]. Indeed, since the side of our sample is only 2 mm long and the indium ohmic contacts are applied by soldering, there is likely a small geometric difference between the xx and yy sides of the sample. While we do not observe any obvious signatures of density gradients in our sample, it is possible that small variations around the mean pressure result in small density fluctuations which may also influence the magnetoresistance.

In Fig. 5.2, we present Hall resistance data in order to showcase further the signatures of the phases throughout the transition. At $P = 6.95$ kbar we find a plateau in R_{xy} quantized to $2h/5e^2$. Such a quantized plateau, when taken together with the minima observed in the longitudinal magnetoresistances shown in Fig. 5.1a, indicates that the ground state at $\nu = 5/2$ is a FQHS at this pressure [4, 5]. This Hall plateau weakens considerably at $P = 7.60$ kbar, indicating that the FQHS at $\nu = 5/2$ approaches an instability. Finally, at $P = 8.26$ kbar we are not able to reliably measure R_{xy} because of the well-known measurement artifact called resistance mixing.

Indeed, as shown in Fig. 5.2c, the resistance peak detected in R_{xx} partially mixes with R_{xy} and produces strong peak-like deviations from the classical value of the Hall resistance at $\nu = 5/2$. This resistance mixing is similar to the well-documented mixing in the nematic phase at $\nu = 9/2$ [7, 8]. In order to mitigate mixing effects, we used the well-known technique of averaging the Hall resistance measured on the same contacts in both positive and negative magnetic fields. Such an averaging, also shown in Fig. 5.2c as the blue trace, reduces the peak-like features at $\nu = 5/2$ and it therefore supports the suggested mixing. However, the cancelation of the peaks is

Fig. 5.2 The Hall resistance at the three representative pressures seen in Fig. 5.1. (**a**) At $P = 6.95$ kbar, there is a quantized Hall plateau at $\nu = 5/2$, signifying a FQHS. (**b**) At $P = 7.60$ kbar, the Hall plateau at $\nu = 5/2$ is weakened, demonstrating a proximity to a critical pressure where the FQHS is nearly destroyed. (**c**) At $P = 8.26$ kbar, there is evidence of mixing from R_{xx}, which is very large. The green and red traces represent the measurement of R_{xy} along the two diagonals of our sample, and the blue trace at $\nu = 5/2$ is the average of the two. This kind of mixing is expected in R_{xy} near a nematic phase. This figure is adapted from Ref. [1]

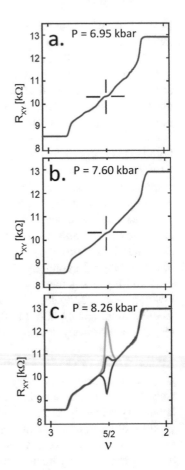

not perfect, presumably because of the different current paths at different orientation of the magnetic field. Nonetheless, quantization at $\nu = 5/2$ is not observed in the nematic phase at $P = 8.26$ kbar.

The evolution of the two longitudinal resistances R_{xx} and R_{yy} is captured over a larger pressure range in the contour plot shown in Fig. 5.3. We focus on the behavior along the line at $\nu = 5/2$. At the lowest measured pressures, the FQHS is shown as a narrow vertical blue line. As the pressure is increased, the $\nu = 5/2$ FQHS weakens, and past a critical pressure, estimated to be $P_c \simeq 7.8 \pm 0.2$ kbar, the nematic phase is stabilized. The nematic phase is seen in Fig. 5.3 as a red island in R_{xx} and as a blue basin in R_{yy}. Our data at $\nu = 5/2$, shown in Fig. 5.3, suggest the possibility of a direct quantum phase transition from a FQHS to the nematic phase as the pressure is tuned through its critical value $P_c^{5/2}$. In Fig. 5.3 the region of stability for the nematic phase is centered near $P \simeq 8.7$ kbar. The nematic phase is weakened by a further increase in pressure until it disappears at an extrapolated value of $P \simeq 10$ kbar. Past this pressure, the resistance does not exhibit a strong anisotropy, thus the ground state past 10 kbar is a rotationally invariant, uniform electron fluid. At $\nu = 5/2$ we

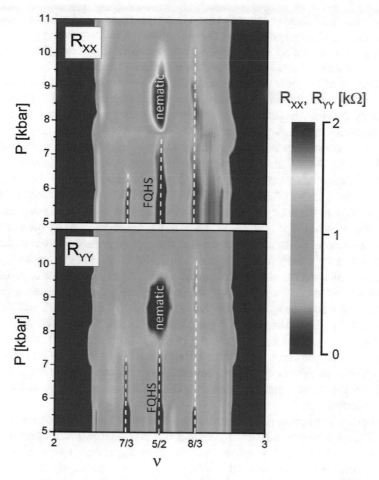

Fig. 5.3 At $\nu = 5/2$, we observe a spatially invariant FQHS at $P < 7.8$ kbar, the nematic phase at $7.8 < P < 10$ kbar, and an isotropic Fermi liquid at $P > 10$ kbar. The nematic phase develops in a narrow range of filling factors $\Delta\nu \simeq 0.15$ centered around $\nu = 5/2$. This figure is adapted from Ref. [1]

find a second quantum phase transition near $P \simeq 10$ kbar between the nematic phase and an isotropic Fermi liquid. We note that in Fig. 5.3 we also see weak FQHSs at $\nu = 7/3$ and 8/3. However, the nematic phase is not stabilized at these filling factors.

5.2   Spontaneous Rotational Symmetry Breaking

In 2DESs with half-filled Landau levels, we differentiate between two types of anisotropies: spontaneous and induced anisotropy. We may draw an analogy between the ground states of the 2DES associated with these anisotropies and the spontaneous and induced magnetism in an interacting spin system. In the absence of an externally applied magnetic field, the spin system exhibits spontaneous symmetry breaking with decreasing temperature, which manifests in a sharp phase transition between the ordered ferromagnet and the disordered paramagnet. In contrast, the development of the ordered phase with the application of an *external magnetic field* is not associated with a thermodynamic singularity. In the 2DES, spontaneous anisotropy develops in the absence of any externally applied symmetry-breaking fields in the nematic phase, at $\nu = 9/2, 11/2, 13/2, 15/2,\ldots$ [7, 8] and at $\nu = 7/2$ [9] at low enough temperatures. As discussed above, however, the ground state at $\nu = 5/2$ was always found to be isotropic in the absence of a symmetry-breaking field [7, 8]. *Induced* anisotropy at $\nu = 5/2$ appears, however, with the application of an external symmetry-breaking field, as discussed in the second chapter [10–14].

In contrast to these experimental results with strain or with in-plane field, the anisotropy we observe at $\nu = 5/2$ in Fig. 5.1c has clearly developed spontaneously. Indeed, because of the hydrostatic nature of the applied pressure, in our experiment the rotational symmetry in the plane of the 2DES is not broken by any external fields. An unintentional in-plane magnetic field may appear in our experiment if the sample tilts inside the cell during the compression process generating the high pressures. However, the isotropic resistance near $\nu = 5/2$ at $P = 6.95$ and $7.60\,$kbar attests that this is not the case. We therefore report a pressure-tuned spontaneous transition at $\nu = 5/2$ from an isotropic FQHS to a quantum Hall nematic phase through an isotropic Fermi liquid phase. Because at $T = 12\,$mK the isotropic liquid is observed in an extremely narrow range of pressures, our data are suggestive of a direct quantum phase transition from the FQHS to the nematic phase in the limit of zero temperatures.

The difference between the spontaneous and induced anisotropic phases at half-filled Landau levels is further highlighted by their contrasting magnetotransport signatures. Although both manifest in anisotropic magnetoresistance, a peculiarity of the spontaneous anisotropy is that it develops over a limited span of filling factors $\Delta\nu \simeq 0.15$ centered on a half-integer filling factor [7, 8]. In contrast, the resistance anisotropy induced by an external in-plane magnetic field at $\nu = 5/2$ is present over a considerably wider range of filling factors $\Delta\nu \simeq 0.6$ (Refs. [10–14]). The observed anisotropy at $P = 8.26\,$kbar shown in Fig. 5.1c, occurring over a narrow range of filling factors $\Delta\nu \simeq 0.15$, is consistent with our earlier conclusion that the ground state at $\nu = 5/2$ is a genuine quantum Hall nematic phase [15, 16] similar to that observed at $\nu = 9/2$ (Refs. [7, 8].)

We note that the orientation of the nematic phase relative to the crystal axes in experiments in the ambient is reproduced in different cooldowns [7, 8]. Similarly, the orientation of the nematic phase at $\nu = 5/2$ observed in the range $7.8 < P < 10$ kbar in our experiment does not change after we change the pressure in our cell at room temperature. In the most general case, one would expect the nematic order to develop along different crystal directions. However, in the GaAs host semiconductor the nematic phase interacts weakly with the host crystal. The origin of this weak interaction is not at present understood [7, 8, 17, 18]. This interaction is, however, responsible for the alignment of the nematic phase with the crystal axes and renders

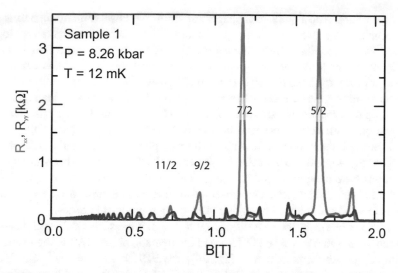

Fig. 5.4 At $P = 8.26$ kbar and higher, nematic phases exist at $\nu = 5/2, 7/2, 9/2$, and $11/2$, though they are highly suppressed at $\nu = 9/2$ and $11/2$. The same hard and easy axes are observed for all nematic phases at all pressures. The data around $\nu = 5/2$ was published in Ref. [1]

the resistance anisotropy readily observable. Using the analogy of the nematic phase with the ferromagnetic phase in interacting spins, in the latter system one expects randomly oriented ferromagnetic domains unless a weak interaction with the crystal field aligns the magnetization of these domains. However, the presence of a weak interaction with the crystal is not required for the nematic state itself to arise. We emphasize that this nematic phase at $\nu = 5/2$ (and as I will discuss in detail later, that which arises at $\nu = 7/2$) is aligned along the same crystalline directions as those at $\nu = 9/2, 11/2$, and so on into higher Landau levels. This can be verified in Fig. 5.4. This plot shows resistance traces at $P = 8.26$ kbar and $T = 12$ mK, at which the nematic phase is stabilized at $\nu = 5/2, 7/2, 9/2$, and $11/2$. The nematic phases in the third Landau level can be seen to be highly suppressed by the application of pressure, likely due to increased influence from disorder in low density samples. Still, at this pressure and at all measured pressures, the hard and easy axes are the same for all nematic phases observed, indicating a similar influence from crystalline fields.

5.3   Topology, Pairing, and the Nematic Phase

The phase transition we have observed is notable in that it involves the change of two different types of order. The collapse of the ordered nematic phase, a traditional Landau phase with broken spatial symmetry [19], is accompanied by the emergence of a topologically ordered phase [20–23] rather than a disordered isotropic phase. The FQHS-to-nematic transition we observe at $\nu = 5/2$ is thus an example of a phase transition which involves the change of both the topological as well as the rotational order across the transition. Such a phase transition was predicted in Ref. [24]. Our observations are incompatible with a direct first order phase transition from the FQHS to the nematic phase, but are compatible with either a direct continuous transition between these two phases or with an intercalation of an isotropic Fermi liquid between these two phases. In the former case we think that the quantum critical point is necessarily described by an exotic theory (not based on the Landau picture) owing to the interplay of the nematic order and the emergent topological order in the non-Abelian FQHS. We note that similar exotic transitions have been proposed in topologically ordered states and in a generalized quantum dimer model [23]. Our observed transition is therefore very special and pushes forward the study of transitions involving topological phases.

We also mention that our transition highlights an interesting relationship between a paired phase—the $\nu = 5/2$ FQHS—and the nematic phase. Within the framework of the composite fermion theory [25, 26], the FQHSs at $\nu = 5/2$ and $7/2$ are due to pairing of composite fermions [24, 27–31]. Pairing and nematicity also appear hand in hand in various other condensed matter systems, namely the high T_c superconductors [32–36]. However, the interplay of nematicity with these paired phases [32, 33] is not understood. The influence of the nematic fluctuations on pairing in the superconductive phase is actively researched [37–41]. Most excitingly, nematic fluctuations may play a role in enhancing pairing [35, 41]. The transition we have found may therefore provide information to theories that attempt to illuminate the relationship of nematic and paired phases.

5.4   Finite Temperature Studies at $\nu = 5/2$

The temperature dependence of the FQHSs and nematic phases at $\nu = 5/2$ represent a crucial set of information for understanding how the phases evolve with pressure. Here I extract relevant temperature scales of the $\nu = 5/2$ FQHS and nematic phase, and construct a summarizing diagram of the phases.

We begin with Fig. 5.5, in which we show the dependence of the longitudinal magnetoresistance for the Landau filling factor range $2 < \nu < 3$ on temperature and pressure. Again, we plot R_{xx} and R_{yy}, this time with magnetic field. Our analysis is focused at $\nu = 5/2$, marked by the large vertical dashed lines in Fig. 5.5. We discuss the different ground states stabilized at $\nu = 5/2$ at different values of pressure. At $P = 6.95$ kbar and $T = 12$ mK, the longitudinal magnetoresistance near $\nu = 5/2$ is vanishingly small and nearly isotropic: the signature of the FQHS at $\nu = 5/2$ [4, 5], identical to the plot in Fig. 5.1a. The density of states of the FQHS at $\nu = 5/2$, similarly to that of any other FQHS, has an energy gap, hence this FQHS is an incompressible quantum liquid [4, 5]. At $P = 7.60$ kbar and $T = 12$ mK, the magnetoresistance at $\nu = 5/2$ remains finite, featureless, and nearly isotropic [1],

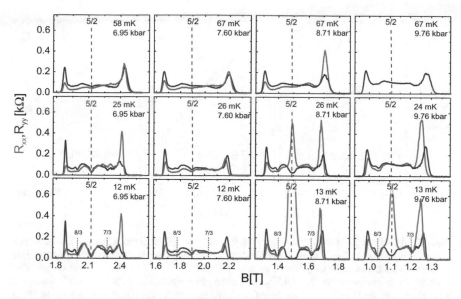

Fig. 5.5 The evolution of magnetotransport between $v = 2$ and 3 at three temperature and four pressure values. The green lines show R_{xx} measured along the $\langle 1\bar{1}0 \rangle$ crystallographic direction of the GaAs host, while the red lines R_{yy} measured along $\langle 110 \rangle$. The longer vertical dashed lines mark $v = 5/2$, while the shorter dotted lines are at $v = 7/3$ and 8/3. The ground state at $v = 5/2$ and at 6.95 kbar is a FQHS, at 7.60 kbar it is a nearly isotropic Fermi fluid, and at 8.71 and 9.76 kbar it is an electronic nematic phase. At 9.76 kbar, the nematic phase is noticeably weaker. Data sets at the lowest temperature for 6.95 and 7.60 kbar are from Ref. [1]. The plots at $P = 6.95, 7.60$, and 8.71 kbar are from Ref. [2]. Reprinted figure with permission from K.A. Schreiber et al., Phys. Rev. B 96, 041107 (2017). Copyright 2017 by the American Physical Society

identical to Fig. 5.1b. Again, we interpret this data at $P = 7.60$ kbar and T $= 12$ mK as evidence for a Fermi-liquid-like state.

In contrast to these, at $P = 8.71$ kbar and T $= 13$ mK, the longitudinal magnetoresistance at $v = 5/2$ is strongly anisotropic. As I have discussed in the previous section, the anisotropic magnetoresistances we observe at $v = 5/2$ and $P = 8.71$ kbar are identical in all aspects to that of the nematic phase forming at $v = 9/2, 11/2$, and so on into higher half-filled Landau levels. Indeed, anisotropy at both of these filling factors develops in the absence of the application of any in-plane B field and in a very limited range of filling factors of width $v \approx 0.15$ around the half-integer value [1, 7–9, 18, 42–45].

In the fourth column, at $P = 9.76$ kbar and $T = 12$ mK, the nematic phase can be seen as well, but the degree of anisotropy is reduced. This is perhaps indicative of the low densities attained at this pressure, meaning the effect of disorder is more influential. The nematic phase is therefore less robust. Indeed, by $T = 25$ mK, the resistances are isotropic at this pressure, and the nematic has been entirely destroyed.

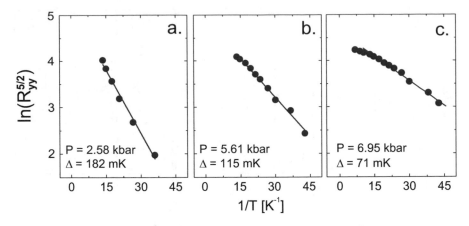

Fig. 5.6 Arrhenius plots using the resistance R_{yy} at $\nu = 5/2$ at three representative pressures. From these we extract the FQHS gap. (**a**) At $P = 2.58$ kbar, the gap is relatively large. (**b**) The gap decreases with the increase of pressure, to $\Delta = 115$ mK at $P = 5.61$ kbar. (**c**) At $P = 6.95$ kbar, we observe the lowest measured FQHS gap of our experiment at $\nu = 5/2$. The data in panels (**a**) and (**c**) are published in Ref. [2]. Reprinted figure with permission from K.A. Schreiber et al., Phys. Rev. B 96, 041107 (2017). Copyright 2017 by the American Physical Society

Magnetoresistance data shown in the lowest row of panels of Fig. 5.5 demonstrate that the ground state at $\nu = 5/2$ as measured near 12 mK evolves from a FQHS toward an electronic nematic phase as the pressure is increased [1]. Figure 5.5 shows how a rising temperature changes the magnetoresistance at $\nu = 5/2$. As a rule, at higher temperatures, features of the magnetoresistance become less pronounced. For example, at $P = 6.95$ kbar, there is an increase of the magnetoresistance at $\nu = 5/2$ as the temperature is raised from 12 to 25 mK. This indicates an enhanced generation of thermally activated excitations in the FQHS. In addition, at $P = 8.71$ kbar, the degree of anisotropy of the nematic phase measured at $T = 26$ mK is weaker than that measured at $T = 13$ mK. At the highest temperature presented here, $T = 67$ mK, the states are nearly destroyed: the FQHS minimum is gone, and the traces are isotropic. Note that only R_{yy} was measured at 67 mK and 9.76 kbar, but we expect it to be isotropic, as it is at 25 mK.

In order to describe the temperature evolution of the observed ground states, we extract a characteristic energy scale associated with them. The FQHS is characterized by the energy gap of the excitations with respect to the ground state, as we have discussed in Chap. 1. The longitudinal magnetoresistance in the presence of an energy gap Δ in the density of states is proportional to $e^{-\Delta/2k_BT}$. Figure 5.6 shows the activated behavior of the FQHS at $\nu = 5/2$ and the extracted energy gaps of the $\nu = 5/2$ FQHS at $P = 2.58$, 5.61, and 6.95 kbar. We find that the energy gap of the $\nu = 5/2$ FQHS decreases with increasing pressure, indicating a weakening of the $\nu = 5/2$ FQHS as the pressure increases.

The temperature dependence of the nematic phase at $\nu = 5/2$ is shown for $P = 8.71$, 9.03, and 9.76 kbar in Fig. 5.7. At relatively high temperatures,

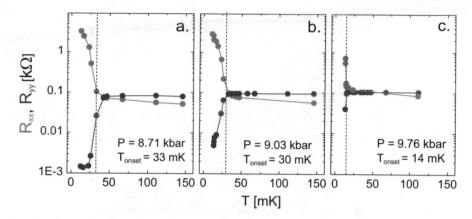

Fig. 5.7 The resistance of the R_{xx} peak and of the R_{yy} minimum as a function of temperature, for the nematic phases at (**a**) $P = 8.71$ kbar, (**b**) $P = 9.03$ kbar, and (**c**) $P = 9.76$ kbar. T_{onset} is shown here as the point at which $R_{xx} = 2R_{yy}$, where the black dashed lines have been placed. Panel (**a**) is from Ref. [2]. Reprinted figure with permission from K.A. Schreiber et al., Phys. Rev. B 96, 041107 (2017). Copyright 2017 by the American Physical Society

the magnetoresistance is nearly isotropic and becomes highly anisotropic as the temperature is decreased. At these high temperatures, we observe a small difference between R_{xx} and R_{yy} which is often seen in experiments and is commonly attributed to imperfections in the sample geometry.

In contrast to the behavior of R_{xx} and R_{yy} at higher temperatures, R_{xx} and R_{yy} sharply deviate from one another at lower temperatures [7, 8, 42, 43]. As seen in Fig. 5.7 the R_{xx}/R_{yy} ratio of the resistances in the two different crystallographic directions exceeds three orders of magnitude at the lowest temperatures. The relatively abrupt onset of anisotropy is a hallmark property for the nematic phase and it defines the onset temperature for nematicity T_{onset}. We estimate T_{onset} by imposing a significant anisotropy $R_{xx} = 2R_{yy}$ in the linearly interpolated data. The dashed line in Fig. 5.7 marks T_{onset} obtained this way.

The dependence on the pressure of the energy gap of the $\nu = 5/2$ FQHS and of the estimated onset temperature of the nematic phase at $\nu = 5/2$ are summarized in Fig. 5.8. We observe that the energy gap of the $\nu = 5/2$ FQHS is monotonically suppressed with increasing pressure. At higher pressures we find that the nematic phase is stabilized at $\nu = 5/2$. In Fig. 5.8, the dashed red line is a guide to the eye for the energy gap of the $\nu = 5/2$ FQHS and the dashed blue line for the onset temperature of the nematic phase at $\nu = 5/2$.

Figure 5.8 can be understood as a phase diagram. Far below the dashed lines the ground state is either a FQHS or a nematic phase. Above the dashed lines there is a Fermi-liquid-like phase. We note that the red dashed line is not a sharp phase boundary, but it represents a crossover between the FQHS and the Fermi liquid. The blue dashed line denotes a transition of an unknown type. The continuous horizontal red line at $T = 0$ indicates the ground state is the $\nu = 5/2$ FQHS, while the

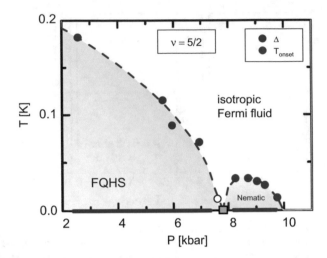

Fig. 5.8 A diagram summarizing the behavior at $\nu = 5/2$ in the $P-T$ phase space. Solid symbols represent the energy gap of the FQHS (red symbols) and the onset temperature of the nematic phase (blue symbols). The open symbol at P = 7.60 kbar and $T = 12$ mK shows that at these parameters we observe a nearly isotropic Fermi fluid. Dashed lines are guides to the eye. The green square is a quantum critical point. This plot is adapted from Ref. [2]. Reprinted figure with permission from K.A. Schreiber et al., Phys. Rev. B 96, 041107 (2017). Copyright 2017 by the American Physical Society

continuous blue line represents the nematic phase in the limit of $T = 0$. Above the dashed lines we have an isotropic Fermi-liquid-like phase. Since data sets at $P = 7.60$ kbar are consistent with a Fermi-liquid-like state, the Fermi liquid is wedged in between the FQHS and the nematic, down to at least 12 mK. The open circle at $P = 7.60$ kbar at $T = 12$ mK in Fig. 5.8 marks this point of the lowest temperature Fermi liquid we accessed. Because the Fermi liquid is wedged in between the two ordered phases, the nematic region forms a dome in the $P - T$ phase diagram.

The phase diagram shown in Fig. 5.8 is an example of an experimentally obtained diagram exhibiting quantum criticality of competing topological and nematic orders. In the vicinity of $P = 7.6$ kbar, this diagram is very similar to the diagram of a quantum phase transition [46]. Earlier we suggested a direct quantum phase transition between these two phases which occurs at the quantum critical point $P_c = 7.8 \pm 0.2$ kbar [1]. This critical point is of an interesting type because one of the phases is topological in nature. As the quantum critical point is crossed with an increasing pressure, the topological order of the FQHS is destroyed while the nematic order is acquired.

Obtaining more detailed data near $P_c^{5/2}$ is quite challenging due to the inability to change the pressure in situ, a limitation of the technique we use. As a result, there are three possible configurations of the phases with respect to one another near the critical pressure, among which we cannot precisely distinguish. These possibilities are presented schematically in Fig. 5.9 in the vicinity of the transition. In Fig. 5.9a, the FQHS and nematic phases at $\nu = 5/2$ overlap, and the phase transition between these phases may be driven directly at finite temperature. In Fig. 5.9b, there are two quantum phase transitions: from the FQHS at $\nu = 5/2$ to the isotropic Fermi fluid, and then from the isotropic Fermi fluid to the nematic phase. The Fermi fluid persists to zero temperature in this picture. Figure 5.9c depicts a single, direct quantum phase transition at the critical pressure P_c from the FQHS to the nematic phase at $\nu = 5/2$. In both panels (b) and (c), any cut in the phase diagram at a finite temperature below the onset of nematic phase will reveal the FQHS, Fermi liquid, and nematic sequence of phases as the pressure is increased. We emphasize that a direct phase transition at $T = 0$ remains the simplest, most elegant interpretation of our data. We think that the phase competition shown in Fig. 5.8 originates from a delicate tuning of the effective electron–electron interaction with pressure, an idea more fully explored in the following chapters.

We note that two different theoretical pictures underlie the two phases involved in this transition, in the following sense. Below the critical pressure, a FQHS requires the existence of composite fermions [25, 47]. In contrast, composite fermions are not required to account for the nematic phase above the critical pressure [15, 16, 48]. The existence of a quantum critical point in Fig. 5.8 thus highlights the dichotomy of the two descriptions of the half-filled Landau level: one based on electrons [15, 16, 48] and another on composite fermions [25, 26, 47].

Finally, we note that the FQHSs developing at $\nu = 7/3$ and $8/3$ deteriorate near the quantum critical point. Indeed, from the data from Fig. 5.5 at $P = 6.95$ and 8.71 kbar, the presence of depressions in the magnetoresistance at $\nu = 7/3$ and $8/3$ at the lowest temperatures reached indicates weak FQHSs at these filling factors. However, at the intermediate pressure $P = 7.60$ kbar and $T = 12$ mK, these weak depressions at $\nu = 7/3$ and $8/3$ have virtually disappeared. In the vicinity of the critical pressure we thus observe a conspicuous loss of electronic correlations responsible for the $\nu = 7/3$ and $8/3$ FQHSs. One possibility is that such a deterioration of the FQHSs at $\nu = 7/3$ and $8/3$ near the quantum critical point could be due to enhanced quantum fluctuations.

5.5   Quantum Phase Transition from Nematic Phase to Fermi Fluid-Like Phase

In Fig. 5.8 there is a second quantum phase transition at high pressures, not depicted on the figure, from the nematic to a Fermi fluid. At these high pressures, nearing 11 kbar, we have attained low electron densities below $5 \times 10^{10} \text{cm}^{-2}$. At such low electron densities we expect that disorder effects do not permit nematic order. We thus think that the destruction of the nematic both at $\nu = 5/2$ and $\nu = 7/2$ at similar electron densities is an indication that disorder became the dominant energy scale at high pressures. This marks a quantum phase transition changing the nematic order to a phase lacking order, even in the limit of zero temperature.

Fig. 5.9 Three possibilities for the FQHS-to-nematic phase transition at $\nu = 5/2$ are depicted schematically here. (**a**) The FQHS phase may intersect with the nematic phase at finite temperature. (**b**) The Fermi fluid may persist to zero temperature, in which case there are two quantum phase transitions: from FQHS to Fermi fluid, and from Fermi fluid to nematic phase. (**c**) There may be a direct quantum phase transition from FQHS to nematic phase at the critical pressure P_c

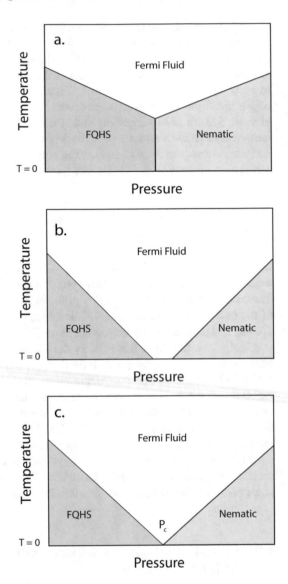

5.6   Conclusion

To conclude, we have observed for the first time a spontaneously arising nematic phase at $\nu = 5/2$ with the application of hydrostatic pressure. The phase transition from FQHS to nematic at $\nu = 5/2$ is an unusual one, changing both the topological order and the nematic order of the phase at $\nu = 5/2$. We have measured the pressure-dependent energy gap of the FQHS at $\nu = 5/2$ and the onset temperature of the nematic phase developing at the same filling factor. These quantities allowed us to map out a summarizing diagram near the instability of the parent Fermi sea toward a FQHS and toward a nematic phase in the $P - T$ parameter space. We found that finite temperature measurements corroborate with the interpretation of a direct quantum phase transition from the FQHS to the nematic phase in the limit of zero temperatures. We have thus demonstrated that the two-dimensional electron gas at $\nu = 5/2$ is a model system which supports competing topological and traditional nematic orders in the $P - T$ parameter space.

References

1. N. Samkharadze, K.A. Schreiber, G.C. Gardner, M.J. Manfra, E. Fradkin, G.A. Csáthy. Nat. Phys. **12**, 191 (2016)
2. K.A. Schreiber, N. Samkharadze, G.C. Gardner, R.R. Biswas, M.J. Manfra, G.A. Csáthy, Phys. Rev. B **96**, 041107 (2017)
3. N. Deng, J.D. Watson, L.P. Rokhinson, M.J. Manfra, G.A. Csáthy, Phys. Rev. B **86**, 201301 (2012)
4. R.L. Willett, J.P. Eisenstein, H.L. Stormer, D.C. Tsui, A.C. Gossard, J.H. English, Phys. Rev. Lett. **59**, 1776 (1987)
5. W. Pan, J.-S. Xia, V. Shvarts, D.E. Adams, H.L. Stormer, D.C. Tsui, L.N. Pfeiffer, K.W. Baldwin, K.W. West, Phys. Rev. Lett. **83**, 3530 (1999)
6. G. Moore, N. Read, Nucl. Phys. B **360**, 362 (1991)
7. M.P. Lilly, K.B. Cooper, J.P. Eisenstein, L.N. Pfeiffer, K.W. West, Phys. Rev. Lett. **82**, 394 (1999)
8. R.R. Du, D.C. Tsui, H.L. Stormer, L.N. Pfeiffer, K.W. Baldwin, K.W. West, Solid State Commun. **109**, 389 (1999)
9. W. Pan, A. Serafin, J.S. Xia, L. Yin, N.S. Sullivan, K.W. Baldwin, K.W. West, L.N. Pfeiffer, D.C. Tsui, Phys. Rev. B. **89**, 241302 (2014)
10. W. Pan, R.R. Du, H.L. Stormer, D.C. Tsui, L.N. Pfeiffer, K.W. Baldwin, K.W. West, Phys. Rev. Lett. **83**, 820 (1999)
11. M.P. Lilly, K.B. Cooper, J.P. Eisenstein, L.N. Pfeiffer, K.W. West, Phys. Rev. Lett. **83**, 824 (1999)
12. B. Friess, V. Umansky, L. Tiemann, K. von Klitzing, J.H. Smet, Phys. Rev. Lett. **113**, 076803 (2014)
13. X. Shi, W. Pan, K.W. Baldwin, K.W. West, L.N. Pfeiffer, D.C. Tsui, Phys. Rev. B **91**, 125308 (2015)
14. J. Xia, V. Cvicek, J.P. Eisenstein, L.N. Pfeiffer, K.W. West, Phys. Rev. Lett. **105**, 176807 (2010)
15. A.A. Koulakov, M.M. Fogler, B.I. Shklovskii, Phys. Rev. Lett. **76**, 499 (1996)
16. R. Moessner, J.T. Chalker, Phys. Rev. B. **54**, 5006 (1996)
17. S.P. Koduvayur, Y. Lyanda-Geller, S. Khlebnikov, G.A. Csáthy, M.J. Manfra, L.N. Pfeiffer, K.W. West, L.P. Rokhinson, Phys. Rev. Lett. **106**, 016804 (2011)
18. J. Pollanen, K.B. Cooper, S. Brandsen, J.P. Eisenstein, L.N. Pfeiffer, K.W. West, Phys. Rev. B. **92**, 115410 (2015)
19. L.D. Landau, E.M. Lifshitz, *Statistical Physics. Part 1*, 3rd edn. Landau and Lifshitz Course of Theoretical Physics, vol. 5. (Elsevier, Amsterdam, 1980)
20. M.Z. Hasan, C.L. Kane, Rev. Mod. Phys. **82**, 3045 (2010)
21. X.L. Qi, S.C. Zhang, Rev. Mod. Phys. **83**, 1057 (2011)
22. Y. Ran, X.G. Wen, Phys. Rev. Lett. **96**, 026802 (2006)
23. E. Ardonne, P. Fendley, E. Fradkin, Ann. Phys. **310**, 493 (2004)
24. E.H. Rezayi, F.D.M. Haldane, Phys. Rev. Lett. **84**, 4685 (2000)

25. J.K. Jain, Phys. Rev. Lett. **63**, 199 (1989)
26. B.I. Halperin, P.A. Lee, N. Read, Phys. Rev. B **47**, 7312 (1993)
27. M. Greiter, X.G. Wen, F. Wilczek, Phys. Rev. Lett. **66**, 3205 (1991)
28. N. Read, D. Green, Phys. Rev. B **61**, 10267 (2000)
29. V.W. Scarola, K. Park, J.K. Jain, Nature **406**, 863 (2000)
30. H. Lu, S. Das Sarma, K. Park, Phys. Rev. B **82**, 201303 (2010)
31. S.A. Parameswaran, S.A. Kivelson, S.L. Sondhi, B.Z. Spivak, Phys. Rev. Lett. **106**, 236801 (2011)
32. B. Keimer, S.A. Kivelson, M.R. Norman, S. Uchida, J. Zaanen, Nature **518**, 179 (2015)
33. E. Fradkin, S.A. Kivelson, J.M. Tranquada, Rev. Mod. Phys. **87**, 457 (2015)
34. V.J. Emery, S.A. Kivelson, J.M. Tranquada, Proc. Natl. Acad. Sci. **96**, 8814 (1999)
35. S.A. Kivelson, E. Fradkin, V.J. Emery. Nature **393**, 550 (1998)
36. S.-W. Cheong, G. Aeppli, T.E. Mason, H. Mook, S.M. Hayden, P.C. Canfield, Z. Fisk, K.N. Clausen, J.L. Martinez, Phys. Rev. Lett. **67**, 1791 (1991)
37. M.A. Metlitski, D.F. Mross, S. Sachdev, T. Senthil, Phys. Rev. B **91**, 115111 (2015)
38. S. Lederer, Y. Schattner, E. Berg, S.A. Kivelson, Phys. Rev. Lett. **114**, 097001 (2015)
39. Y. Schattner, S. Lederer, S.A. Kivelson, E. Berg. Phys. Rev. X **6**, 031028 (2016)
40. P.T. Dumitrescu, M. Serbyn, R.T. Scalettar, A. Vishwanath, Phys. Rev. B **94**, 155127 (2016)
41. A. Mesaros, M.J. Lawler, E.-A. Kim, Phys. Rev. B **95**, 125127 (2017)
42. G. Sambandamurthy, R.M. Lewis, H. Zhu, Y.P. Chen, L.W. Engel, D.C. Tsui, L.N. Pfeiffer, K.W. West, Phys. Rev. Lett. **100**, 256801 (2008)
43. M.E. Msall, W. Dietsche, New J. Phys. 17, 043042 (2015)
44. Q. Shi, M.A. Zudov, J.D. Watson, G.C. Gardner, M.J. Manfra, Phys. Rev. B **93**, 121404 (2016)
45. Q. Shi, M.A. Zudov, J.D. Watson, G.C. Gardner, M.J. Manfra, Phys. Rev. B **93**, 121411 (2016)
46. S. Sachdev, *Quantum Phase Transitions*, 2nd edn. (Cambridge University Press, Cambridge, 2011)
47. J.K. Jain, *Composite Fermions* (Cambridge University Press, Cambridge, 2007)
48. E. Fradkin, S.A. Kivelson, Phys. Rev. B. **59**, 8065 (1999)

Chapter 6
Universality of the Fractional Quantum Hall State-to-Nematic Phase Transition at Half-Filling in the Second Landau Level

We have demonstrated that the FQHS at filling factor $\nu = 5/2$ has a proximity to a nematic phase. Indeed, pressure drives what appears to be a quantum phase transition from FQHS to nematic phase. Because hydrostatic pressure preserves rotational symmetry, this observation raises many questions about the mechanism of the transition. In the discussion of Fig. 5.4, I alluded to the fact that a nematic phase also arises at $\nu = 7/2$, a cousin of $\nu = 5/2$ that is expected to share the same physics. One avenue, therefore, toward illuminating the FQHS-to-nematic transition at $\nu = 5/2$ is to study the FQHS-to-nematic transition that occurs at $\nu = 7/2$. The appearance of the FQHS-to-nematic phase transition at both half-filled spin branches in the second Landau level emphasizes the special nature of this Landau level. Analyzing the temperature dependence of the phases at $\nu = 7/2$, we find that the FQHS-to-nematic phase transition appears to be a quantum phase transition like that at $\nu = 5/2$. Comparing the two filling factors, we begin to answer questions about the role that pressure plays in driving the transition.

6.1 Observation of the FQHS-to-Nematic Phase Transition at $\nu = 7/2$

For this set of experiments, we measured a second sample, sample 2, under hydrostatic pressure. We were unable to extract detailed FQHS measurements at $\nu = 7/2$ in sample 1. The measurement is very time-consuming, so we first focused on $\nu = 5/2$ and the nematic phase at $\nu = 7/2$ in sample 1. However, due to repeated thermal cycling at high pressure, the feedthrough was broken before we returned to low pressure to study the $\nu = 7/2$ FQHS. Sample 1 was lost in this process, so we continued with sample 2. Sample 2 is a 30 nm quantum well sample with an as-grown density of $29.0 \times 10^{10}\,\mathrm{cm^{-2}}$, cut to a $2 \times 2\,\mathrm{mm^2}$ square and annealed

© Springer Nature Switzerland AG 2019
K. A. Schreiber, *Ground States of the Two-Dimensional Electron System at
Half-Filling under Hydrostatic Pressure*, Springer Theses,
https://doi.org/10.1007/978-3-030-26322-5_6

with indium/tin contacts. It is very similar in structure to sample 1, but is cut from a different wafer. As in the measurement of sample 1, measurements were performed in a dilution refrigerator, using a standard low frequency lock-in technique. The magnetic field up to 10 T was applied perpendicularly to the plane of the electron gas. Before cooling to low temperatures, samples were illuminated at 10 K using a red light emitting diode. Again, we estimate the lowest electronic temperatures reached in this pressure cell are about 12 mK. The sample was pressurized using the same techniques previously described. We focus our study on the second Landau level over a wide pressure range. To review the filling factors found in the two spin branches, the second orbital Landau level in GaAs corresponds to the $2 < \nu < 4$ range. Of this range, the $2 < \nu < 3$ is the lower spin branch, while the $3 < \nu < 4$ range is the upper spin branch. Therefore at $\nu = 5/2$ and $\nu = 7/2$ the system has half-filled Landau levels with the same orbital quantum number, but different spin quantum numbers.

Figure 6.1 highlights the evolution of the magnetoresistance in the two spin branches of the second orbital Landau level at the lowest temperature of $T \approx 12$ mK reached in our pressure cell. Traces are measured along two mutually perpendicular directions as in sample 1: R_{xx} along the $\langle 1\bar{1}0 \rangle$ crystallographic direction of GaAs, and R_{yy} along the $\langle 110 \rangle$ direction. These traces show several features which can be associated with known ground states of the electron gas at ambient pressure [1, 2]; in the following we focus our attention to $\nu = 5/2$ and $\nu = 7/2$. The magnetoresistance at $\nu = 5/2$ is isotropic and vanishing at 3.26 and 7.22 kbar, signaling a FQHS [3, 4]. The magnetoresistance at $\nu = 5/2$ is strongly anisotropic at 9.26 kbar and remains slightly anisotropic at 10.54 kbar, exhibiting therefore nematic behavior [5, 6] at these pressures. This behavior with increasing pressure is consistent with a FQHS, quantum Hall nematic, isotropic Fermi fluid sequence of ground states [7].

The magnetoresistance trend at $\nu = 7/2$ shown in Fig. 6.1 is qualitatively similar to that at $\nu = 5/2$ as it evolves from isotropic and nearly vanishing at 3.26 kbar, to strongly anisotropic at 7.22 and 9.26 kbar, to weakly anisotropic at 10.54 kbar. This behavior at $\nu = 7/2$ suggests the same sequence of ground states as at $\nu = 5/2$ and hints at the existence of a FQHS-to-nematic transition at $\nu = 7/2$.

At certain pressures, Fig. 6.1 shows the same type of ground states at both $\nu = 5/2$ and $7/2$. Indeed, at $P = 3.26$ kbar we observe two FQHSs, while at $P = 9.26$ and 10.54 kbar we observe two nematic phases. This arrangement of similar ground states at different half-filled spin branches of a given orbital Landau level is typical for samples in the ambient. For example, ground states at both $\nu = 5/2$ and $7/2$ in the second Landau level are FQHSs [8] and those at $\nu = 9/2$ and $11/2$ in the third Landau level are nematic states [5, 6]. In Fig. 6.1b we observe an exception to such an arrangement. Indeed, at $P = 7.22$ kbar, the ground state at $\nu = 5/2$ is a FQHS, while that at $\nu = 7/2$ is the nematic. This asymmetry implies that the nematic at $\nu = 7/2$ is stabilized at a lower pressure than that at $\nu = 5/2$.

Enhanced quantum fluctuations may have observable consequences close to the critical point. A recent theory has examined the influence of the nematic fluctuations on the paired FQHS [9]. Nematic fluctuations may also influence the nematic

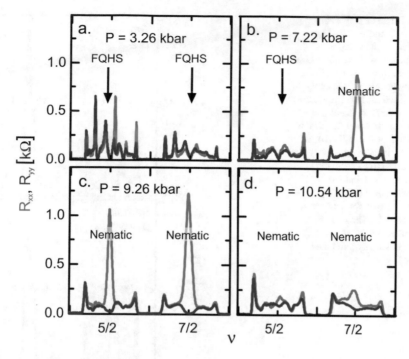

Fig. 6.1 The development of the nematic phases with the application of pressure in sample 2 at base temperature $T \approx 12$ mK. We progress, with increasing pressure, from (**a**) a FQHS at both $\nu = 5/2$ and $\nu = 7/2$, (**b**) a FQHS at $\nu = 5/2$ and a nematic phase at $\nu = 7/2$, (**c**) nematic phases at both $\nu = 5/2$ and $\nu = 7/2$, (**d**) nearly destroyed nematic phases at both filling factors at high pressure

phase itself in a description beyond the mean field [10, 11]. Our data show several anomalies close to the quantum critical point which may be related to fluctuation effects. One anomaly, shown in Fig. 6.1c, is that the resistance anisotropy at $\nu = 7/2$ exceeds that at $5/2$. At fixed density and fixed temperature, a larger anisotropy typically develops in the lower spin branch. For example, in the third orbital Landau level the anisotropy observed at $\nu = 9/2$ is larger than that at $\nu = 11/2$ [5, 6]. These effects may merit further investigation.

It is worth revisiting sample 1, the sample studied in the previous chapter and in references [7] and [12], to compare the ranges of pressure in which the nematic phase is stabilized in these samples. As mentioned, the FQHS at $\nu = 7/2$ was unfortunately not observed in detail in sample 1. Fairly detailed temperature dependence data of the nematic phase at $\nu = 7/2$ was, on the other hand, acquired in this sample. Figure 6.2a, b displays two representative traces of $\nu = 5/2$ and $\nu = 7/2$. In Fig. 6.2a, at $P = 5.96$ kbar, we see we have attained the nematic phase at $\nu = 7/2$. However, as in Fig. 6.1b, the FQHS still exists at $\nu = 5/2$.

Increasing the pressure beyond the critical point of the transition at $\nu = 5/2$, we drive the transition to $\nu = 5/2$ as well, seen in Fig. 6.2b. It is unequivocal,

Fig. 6.2 The states at $\nu = 5/2$ and $\nu = 7/2$ in sample 1, the sample described in [7, 12], at a base temperature T $= 12$ mK. (**a**) As in the sample presently described, at lower pressures, even when a nematic phase develops at $\nu = 7/2$, there is a FQHS at $\nu = 5/2$. (**b**) At higher pressures, we drive the transition to the nematic at both $\nu = 5/2$ and $\nu = 7/2$. Detailed data of the FQHS at $\nu = 7/2$ was not obtained in this sample. The nematic phase data around $\nu = 5/2$ was previously published in [7]

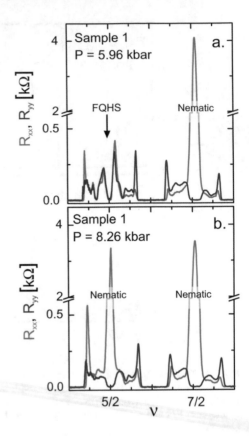

then, that pressure affects the upper and lower spin branches of the second Landau level differently, and the critical pressure is not the same for the FQHS-to-nematic transitions at $\nu = 5/2$ and $\nu = 7/2$.

One will note that the peak resistance R_{xx} in the nematic phase in this sample is four times higher than that observed in sample 2. The reasons for this are not known. The electron temperature may not have been as cold in the second pressurization campaign, or the sample quality simply may have been not conducive to such high resistances, whether due to illumination issues or otherwise.

One more anomaly is worth noting upon comparison of Figs. 6.1 and 6.2. The transitions do not occur at the same pressure at $\nu = 5/2$ in the two samples. Likewise at $\nu = 7/2$, the critical pressure is not the same in the two samples. This in fact has a trivial explanation: different ambient pressure densities in the two samples. Sample 2 has the slightly higher ambient density of 2.9×10^{11} cm^{-2}. This merely means a higher pressure is needed to reduce the density to the same value attained in sample 1, which has ambient pressure density 2.8×10^{11} cm^{-2}. Taken together, these facts provide evidence that pressure is not the primary driver of the transition.

For completeness, we demonstrate that the observed FQHSs are indeed well quantized, with plateaus in the Hall resistance. Figure 6.3 shows the quantized Hall plateaus of $\nu = 5/2$ and $\nu = 7/2$ in sample 2 at 12 mK. Panels (a) and (b) depict

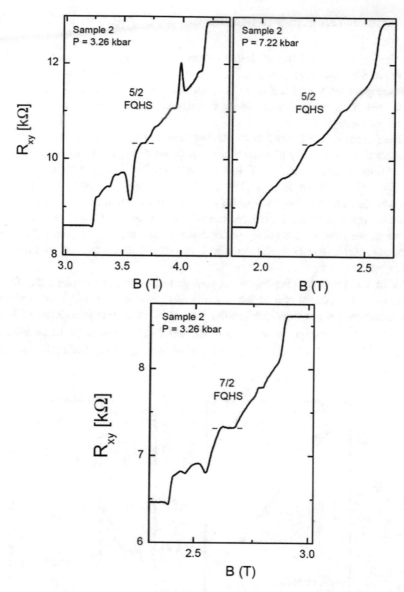

Fig. 6.3 The Hall resistance at two pressures in the pressurized sample 2, showing the quantized resistance of the FQHSs at $\nu = 5/2$ and $\nu = 7/2$. The top two panels show the region of filling factors around $\nu = 5/2$ at $P = 3.26$ and $P = 7.22$ kbar, corresponding to Fig. 6.1a, b above. Panel (c) shows the region of filling factors around $\nu = 7/2$, corresponding to Fig. 6.1a

the evolution of $\nu = 5/2$ with pressure, at $P = 3.26$ kbar and $P = 7.22$ kbar, respectively. These correspond to the FQHSs at $\nu = 5/2$ seen in Fig. 6.1a, b. Panel (c) depicts the Hall resistance of $\nu = 7/2$ at $P = 7.22$ kbar, corresponding to the $\nu = 7/2$ FQHS seen in Fig. 6.1a. Again, this shows that $\nu = 7/2$ at this pressure is a FQHS.

6.2   Finite Temperature Studies at $v = 5/2$ and $v = 7/2$

In order to understand the evolution of phases with pressure, we turn to finite temperature measurements. We define the onset temperature for the nematic T_{onset} as the temperature at which $R_{xx} = 2R_{yy}$ and the energy gap Δ of a FQHS by fitting the magnetoresistance to an activated expression $e^{-\Delta/2k_B T}$, just as we have done in the study of sample 1.

These temperature dependences can be seen in Fig. 6.4 for two representative pressures at $v = 7/2$. By plotting these two quantities against pressure, we obtain the stability diagrams in $P - T$ space shown in Fig. 6.5. The stability diagram at $v = 5/2$ has three regions [12]. At low pressures, we observe a fractional quantum Hall ground state that possesses thermally excited quasiparticles at finite T; the energy gap of the FQHS decreases with an increasing pressure. At higher pressures, we observe nematicity under a dome-like region. At even higher pressures, approaching 11 kbar, the nematic is destroyed into a featureless Fermi-like fluid.

As in the previous chapter, we argue that the simplest explanation for the sequence of the phases and of the stability diagram at $v = 5/2$ is the existence of two quantum phase transitions in the limit of $T = 0$: one from a paired FQHS to the nematic occurring at $P_c^{5/2}$, and another from the nematic to an isotropic Fermi fluid at $\tilde{P}_c^{5/2}$ [7, 12]. Figure 6.5a demonstrates that this earlier result at

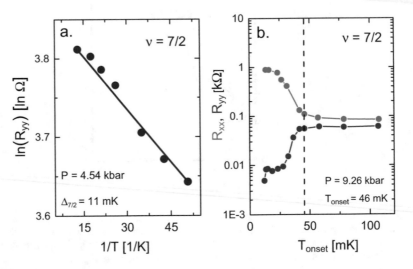

Fig. 6.4 (**a**) The gap of a FQHS at $v = 7/2$ in sample 2 at a representative pressure, $P = 4.54$ kbar. (**b**) The onset temperature of a nematic phase at $v = 7/2$, at a representative pressure $P = 9.26$ kbar

Fig. 6.5 The FQHS gap and the nematic phase onset temperature at (**a**) $\nu = 5/2$ and (**b**) at $\nu = 7/2$ in sample 2 plotted with pressure. The FQHS gap (red circles) decreases and appears to close. The nematic phase appears after the gap closes (blue circles). The green squares represent the extrapolated critical points of the FQHS-to-nematic transition at zero temperature, $P_c^{5/2}$ and $P_c^{7/2}$. The orange squares represent the extrapolated critical points of the transition from nematic to disordered Fermi-like fluid, $\tilde{P}_c^{5/2}$, and $\tilde{P}_c^{7/2}$. Data from Ref. [13]

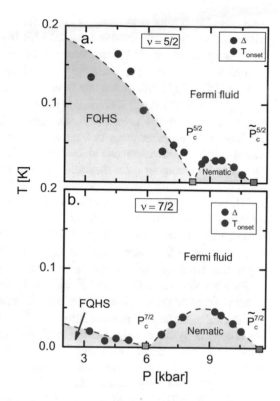

$\nu = 5/2$ is reproducible in a sample of similar structure and of similar density, but cut from a different wafer [12]. Furthermore, the stability diagram at $\nu = 7/2$, shown in Fig. 6.5b, is qualitatively similar to that at $\nu = 5/2$ as it also exhibits the same phases and the same two quantum critical points. Not pictured here are two points measured at pressures 5.78 kbar and 6.47 kbar at which we do not observe a measurable gap at $\nu = 7/2$, but at which $\nu = 7/2$ is isotropic down to our lowest measured temperature of 12 mK. These are analogous to the state seen at $\nu = 5/2$ in Fig. 5.1b.

Our observation of competition of the FQHS and the nematic near the quantum critical point highlights the importance of pairing in our experiments. Of the large number of FQHSs forming in the second Landau level [1–4, 8] only the paired FQHSs at $\nu = 5/2$ and $7/2$ show the pressure induced transition to the nematic. Indeed, the nematic in our pressurized samples does not develop at well-known filling factors, such as the ones at $\nu = 7/3$, $8/3$, $11/5$, or $14/5$, at which the ground state in the ambient are FQHSs lacking pairing. Taken together, these results establish the universal nature of the stability diagram and of the paired FQHS-to-nematic quantum phase transition in the second orbital Landau level.

We estimate the critical pressure of the FQHS-to-nematic transition to be halfway between the highest pressure for the FQHS and the lowest pressure for the nematic. We obtain $P_c^{5/2} = 8.2 \pm 0.5$ kbar and $P_c^{7/2} = 5.9 \pm 0.6$ kbar; these critical points are marked in Fig. 6.5 by green squares. The critical pressure at $\nu = 5/2$ is consistent with 7.8 kbar, the value found in sample 1 [7, 12]. Again, we attribute the difference of the two pressures to the 3% difference in the as-grown density of the two samples.

Strikingly, the critical pressure $P_c^{7/2} = 5.9$ kbar at $\nu = 7/2$ is much reduced from its value at $\nu = 5/2$. We notice that in our sample the ratio of the critical pressures $P_c^{5/2} / P_c^{7/2} = 8.2/5.9 \approx 1.4$ is equal to the ratio of the two filling factors $7/5 = 1.4$. This result suggests that pressure is not a primary driving parameter of the transition, but there may be other ways to induce the same quantum phase transition. This hypothesis is not unreasonable since pressure tunes all band parameters [14–19]. The quantity changing most dramatically with pressure is the electron density: it decreases linearly with pressure to nearly 20% of its value in the ambient at 10 kbar [7, 12, 14]. In Fig. 6.6 we explore the premise of other driving parameters by plotting the nematic onset temperature against pressure, electron density, and magnetic field. Figure 6.6c is particularly significant, showing that in our sample the critical point is at nearly the same magnetic field: $B_c^{5/2} = 1.91$ T and $B_c^{7/2} = 1.94$ T. This is suggestive that magnetic field has an important role to play in driving the transition.

We note that as in the study of sample 1, there is a phase transition at high pressure from nematic phase to a Fermi-like liquid phase dominated by disorder. The critical pressures of this transition, $\tilde{P}_c^{5/2} = 11.0$ kbar and $\tilde{P}_c^{7/2} = 11.4$ kbar, are estimated by linear extrapolation to $T = 0$ of the nematic onset temperatures forming at the two highest pressures. These critical points are marked in Fig. 6.5 by orange squares. When comparing the critical values of different parameters at $\nu = 5/2$ and $7/2$ which may drive the nematic-to-Fermi fluid transition we find that, in contrast to the FQHS-to-nematic transition, this transition occurs at nearly the same pressure, at values of the electron density close to each other $\tilde{n}_c^{5/2} = 5.2 \times 10^{10}$ cm^{-2} and $\tilde{n}_c^{7/2} = 4.5 \times 10^{10}$ cm^{-2}, but at very different magnetic fields. The nematic onset temperature as function of these parameters is seen in Fig. 6.6.

A comparison of the onset temperatures in sample 1 and 2 as functions of magnetic field provides valuable insight as well. Figure 6.7 shows a plot of the onset temperatures in the two samples at $\nu = 5/2$ and $\nu = 7/2$. The black points are the onset temperatures of the nematic phase at $\nu = 7/2$, and the blue are those of the nematic phase at $\nu = 5/2$. The dashed lines and open symbols correspond to sample 1, and the solid lines and symbols correspond to sample 2. The qualitative and quantitative similarities are immediately apparent. The onset temperatures of the nematic phases in the two samples are in good agreement. The nematic phase at $\nu = 7/2$ is generally more robust to temperature than that at $\nu = 5/2$, attaining maximum onset temperatures of $T \approx 45$ mK in both samples. The nematic phase at $\nu = 7/2$ is also stabilized in the magnetic field range 0.7–1.9 T for both samples. Likewise, in both samples the nematic phase at $\nu = 5/2$ achieves a maximum onset temperature near $T \approx 30$ mK, stabilized between 1.0–1.9 T. The nematic phase appearance at $\nu = 5/2$ is quite abrupt for both samples, near 1.9 T. The appearance for $\nu = 7/2$ is more gradual, but still begins near 1.9 T. We do not estimate critical magnetic fields here for sample 1, due to the lack of FQHS data taken at $\nu = 7/2$, but as in sample 2, it is clear that the critical magnetic field must be near 1.9 T for both

Fig. 6.6 The onset temperatures of the nematic phases at $\nu = 5/2$ (open circles) and $\nu = 7/2$ (closed circles) in sample 2 as functions of (**a**) pressure, (**b**) electron density, and (**c**) magnetic field. The green squares represent the extrapolated critical points of the FQHS-to-nematic transitions, and the orange squares represent the extrapolated critical points of the nematic-to-Fermi liquid transition. The lines are guides to the eye. Data from Ref. [13]

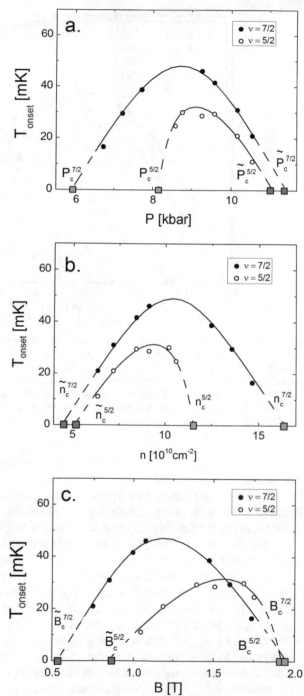

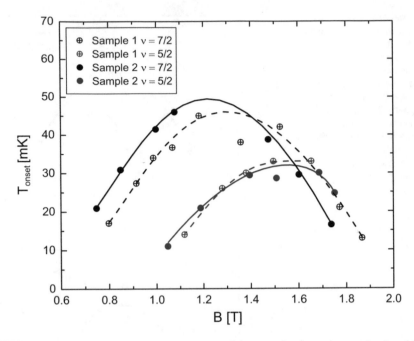

Fig. 6.7 A comparison of the onset temperatures of the nematic phases in samples 1 and 2 at $\nu = 7/2$ and $\nu = 5/2$. The black points are the onset temperatures of the nematic phase at $\nu = 7/2$, and the blue are those of the nematic phase at $\nu = 5/2$ The dashed lines and open symbols correspond to sample 1, and the solid lines and symbols correspond to sample 2. The onset temperature plotted with magnetic field. Notice that the magnetic field near the transition from FQHS to nematic phase in both samples, and at both filling factors, is near $B \approx 1.9\,T$. Data from Ref. [13]

filling factors. This similar behavior in the two samples provides strong evidence that the transition is not sample dependent. It also provides strong evidence for our idea that $B \approx 1.9\,T$ may be a universal, critical magnetic field for triggering the FQHS-to-nematic phase transition in the second Landau level.

Of interest is the fact that the nematic phase at $\nu = 7/2$ is generally more robust to thermal excitations and persists over a wider density range than that at $\nu = 5/2$, despite the fact that the FQHS gap at $\nu = 7/2$ is nearly always smaller than that of $\nu = 5/2$. It may be that enhanced pairing correlations at $\nu = 5/2$, compared to those at $\nu = 7/2$, compete with and weaken the nematic phase at $\nu = 5/2$. This observation invites investigation into the relative strength of the orders at play.

Finally, we provide a further demonstration that the nematic phase observed at $\nu = 5/2$ and $\nu = 7/2$ here is spontaneously arising, because we were able to compare this nematic phase with tilt-induced anisotropy in sample 2. We obtained magnetotransport data for this sample tilted to an approximate angle of $\theta = 35°$. Two experimental signatures provide evidence for a tilted sample. The first is that anisotropy observed in longitudinal resistance arises over a broader range of filling

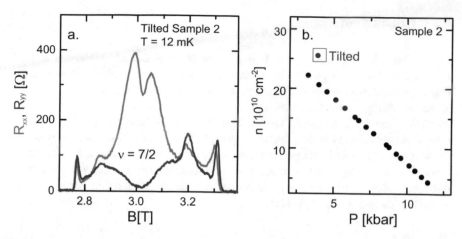

Fig. 6.8 (a) Longitudinal resistance traces around $v = 7/2$ in sample 2 tilted to an estimated angle of 35°, with R_{xx} in green and R_{yy} in red. The pressure on the sample is 3.9 kbar. The anisotropy in the resistance is due to in-plane magnetic field. (**b**)Tilting the sample overestimates the electron density. The apparent density measured at this pressure is depicted by the red point, well above the line of the expected density decrease with pressure (black points)

factor than that which arises in the spontaneous nematic phase, and weak local minima may still be observed in the resistance peak that forms. Such a trace can be observed in Fig. 6.8a, showing that anisotropy appears throughout the upper spin branch of the second Landau level. A dip in the peak, near $v = 7/2$, is evident, and indeed is a frequently observed signature of anisotropy induced in tilted samples [20–24].

Secondly, even more concrete evidence for a tilted sample can be obtained from careful observation of the expected decrease of electron density with pressure. We have observed that density should decrease at a rate of about 2.2×10^{10} cm^{-2}/kbar. However, in a tilted sample, the density will appear *higher* than it is expected to be. Density may be correctly obtained from the perpendicularly applied magnetic field at a known filling factor: $n_{actual} \propto B_{\perp}$. However, if the sample is unknowingly tilted, the perpendicularly applied magnetic field is less than the total applied magnetic field: $B_{\perp} = B_{total} \cos \theta$. The total magnetic field is what is experimentally measured, so in the case of a tilted sample, we measure an incorrect apparent density: $n_{apparent} \propto B_{total} = B_{\perp} / \cos \theta$. Therefore, the incorrectly measured density $n_{apparent}$ will always overestimate the correct density n_{actual}: $n_{apparent} = n_{actual} / \cos \theta$. Such an overestimated density is seen in Fig. 6.8b as the red point. It marks a large deviation from the expected curve of density's decrease with pressure. In this manner, inadvertent tilting can be detected if the measured density is much larger than predicted at a given pressure. When we observed that this sample had tilted, prior to obtaining any of the above presented data, the pressure cell feedthrough was opened, and the tilt was corrected.

6.3   Conclusion

In summary, the transition from FQHS to nematic phase was observed at *both* $\nu = 5/2$ and $\nu = 7/2$ under hydrostatic pressure in *two* similarly grown 30 nm GaAs quantum well samples. Although this transition occurs at different pressures at the two filling factors in the two samples, the transition occurs at almost the same *magnetic field* for both filling factors in both samples. The apparently important role of magnetic field invites further exploration. As we will discuss in detail in the analysis of the following chapter, the magnetic field at which a FQHS occurs determines the magnitude of electron–electron interactions. Therefore, we have obtained a hint of the important role of electron–electron interactions have to play in the FQHS-to-nematic transition.

References

1. J.S. Xia, W. Pan, C.L. Vicente, E.D. Adams, N.S. Sullivan, H.L. Stormer, D.C. Tsui, L.N. Pfeiffer, K.W. Baldwin, K.W. West, Phys. Rev. Lett. **93**, 176809 (2004)
2. E. Kleinbaum, A. Kumar, L.N. Pfeiffer, K.W. West, G.A. Csáthy, Phys. Rev. Lett. **114**, 076801 (2015)
3. R.L. Willett, J.P. Eisenstein, H.L. Stormer, D.C. Tsui, A.C. Gossard, J.H. English, Phys. Rev. Lett. **59**, 1776 (1987)
4. W. Pan, J.-S. Xia, V. Shvarts, D.E. Adams, H.L. Stormer, D.C. Tsui, L.N. Pfeiffer, K.W. Baldwin, K.W. West, Phys. Rev. Lett. **83**, 3530 (1999)
5. M.P. Lilly, K.B. Cooper, J.P. Eisenstein, L.N. Pfeiffer, K.W. West, Phys. Rev. Lett. **82**, 394 (1999)
6. R.R. Du, D.C. Tsui, H.L. Stormer, L.N. Pfeiffer, K.W. Baldwin, K.W. West, Solid State Commun. **109**, 389 (1999)
7. N. Samkharadze, K.A. Schreiber, G.C. Gardner, M.J. Manfra, E. Fradkin, G.A. Csáthy, Nat. Phys. **12**, 191 (2016)
8. J.P. Eisenstein, K.B. Cooper, L.N. Pfeiffer, K.W. West, Phys. Rev. Lett. **88**, 076801 (2002)
9. A. Mesaros, M.J. Lawler, E.-A. Kim, Phys. Rev. B **95**, 125127 (2017)
10. A.A. Koulakov, M.M. Fogler, B.I. Shklovskii, Phys. Rev. Lett. **76**, 499 (1996)
11. R. Moessner, J.T. Chalker, Phys. Rev. B. **54**, 5006 (1996)
12. K.A. Schreiber, N. Samkharadze, G.C. Gardner, R.R. Biswas, M.J. Manfra, G.A. Csáthy, Phys. Rev. B **96**, 041107 (2017)
13. K.A. Schreiber, N. Samkharadze, G.C. Gardner, Y. Lyanda-Geller, M.J. Manfra, L.N. Pfeiffer, K.W. West, G.A. Csáthy, Nat. Commun. **9**, 2400 (2018)
14. D.K. Maude, J.C. Portal, Parallel transport in low-dimensional semiconductor structures. *Semiconductors and Semimetals*, vol. 55 (Elsevier, Amsterdam, 1998)
15. Z. Wasilewski, R.A. Stradling, Semicond. Sci. Technol. **1**, 264 (1986)
16. S. Holmes, D.K. Maude, M.L. Williams, J.J. Harris, J.C. Portal, K.W.J. Barnham, C.T. Foxon, Semicond. Sci. Technol. **9**, 1549 (1994)
17. L. Dmowski, J.C. Portal, Semicond. Sci. Technol. **4**, 211 (1989)
18. G. Gregoris, D. Lavielle, J. Beerens, S. Ben Amor, J.C. Portal, F. Alexandre, Semicond. Sci. Technol. **4**, 317 (1989)
19. J. Beerens, G. Gregoris, J.C. Portal, J.L. Robert, J.M. Mercy, F. Alexandre, Semicond. Sci. Technol. **3**, 577 (1988)
20. J. Xia, J.P. Eisenstein, L.N. Pfeiffer, K.W. West, Nat. Phys. **7**, 845 (2011)

21. Y. Liu, S. Hasdemir, M. Shayegan, L.N. Pfeiffer, K.W. West, K.W. Baldwin, Phys. Rev. B **88**, 035307 (2013)
22. Q. Shi, M.A. Zudov, J.D. Watson, G.C. Gardner, M.J. Manfra, Phys. Rev. B **93**, 121404 (2016)
23. Q. Shi, M.A. Zudov, J.D. Watson, G.C. Gardner, M.J. Manfra, Phys. Rev. B **93**, 121411 (2016)
24. B. Friess, V. Umansky, L. Tiemann, K. von Klitzing, J.H. Smet, Phys. Rev. Lett. **113**, 076803 (2014)

Chapter 7
Origin of the Fractional Quantum Hall State-to-Nematic Phase Transition in the Second Landau Level

We now discuss possible origins for the observed isotropic FQHS-to-nematic phase transition. As we have seen in the previous chapter, the transitions for $\nu = 5/2$ and $\nu = 7/2$ occur not at the same pressure, but the same magnetic field. We have driven $\nu = 5/2$ and $\nu = 7/2$ to this critical magnetic field by decreasing the electron density, one of the effects of applying pressure. Interestingly, such low density electron samples have been measured at ambient pressure before, with no sign of the nematic phase at $\nu = 5/2$ and only incipient anisotropy at $\nu = 7/2$ [1]. We therefore address our observation by considering the effect not simply of magnetic field, but of the magnitude of electron–electron interactions attained in our pressurized sample. In this chapter, I analyze the electron–electron interaction parameters attained in our experiment. We find that the nematic phases are stabilized within the same regimes of these electron–electron interaction parameters at both $\nu = 5/2$ and $\nu = 7/2$. Motivated by this finding, we study a sample at *ambient* pressure, grown so that the degree of electron–electron interactions at $\nu = 7/2$ is the same as that attained in the pressurized sample. Excitingly, we also observe a nematic phase at $\nu = 7/2$ in this unpressurized sample.

7.1 Tuning the Electron–Electron Interactions with Landau Level Mixing

The problem of understanding a FQHS in high Landau levels is a difficult one. Extensive theory work has pushed forward the understanding of the electron–electron interactions in realistic samples. Recent works account for the fact that the electron system is not purely 2D, but does have a finite width. Additionally, ground states in higher Landau levels feel the effect of hybridization from neighboring Landau levels, complicating the problem further. Two parameters encode the

© Springer Nature Switzerland AG 2019
K. A. Schreiber, *Ground States of the Two-Dimensional Electron System at Half-Filling under Hydrostatic Pressure*, Springer Theses,
https://doi.org/10.1007/978-3-030-26322-5_7

electron–electron interactions in such a system: Landau level mixing (LLM) and finite well width. As the ability of experiment to access more and more fragile and complex ground states improves, both LLM and the width of the quantum well must be considered carefully.

When Landau level mixing is neglected, the problem of a FQHS in an excited Landau level is projected onto the lowest Landau level. Doing this neglects the influence from lower filled Landau levels or from empty higher Landau levels. This turns out to be a valid approximation as long as the Landau level mixing parameter, defined as

$$\kappa = \frac{E_{Coulomb}}{E_{Cyclotron}} = \frac{e^2/4\pi\epsilon l_B}{\hbar eB/m} \tag{7.1}$$

is small. Here, $l_B = \sqrt{h/eB}$ is the magnetic length. When κ is not small, fluctuations occurring in the surrounding totally empty and full Landau levels must be accounted for, because they will play a role in determining the ground state of certain filling factors, especially in higher Landau levels [2].

At $\nu = 5/2$, the role of Landau level mixing is especially important, because different values of these parameters are expected to stabilize different ground states at that filling factor [2–7]. For example, in the limit of $\kappa = 0$, the Pfaffian and Anti-Pfaffian wavefunctions are degenerate. Taking LLM into account lifts the degeneracy between these two states, and many recent results in fact favor the Anti-Pfaffian over the Pfaffian as the $\nu = 5/2$ ground state under LLM [3–5]. In practice, LLM can be changed by changing the electron density. Changing electron density changes the magnetic field at which a given filling factor occurs according to $B \propto n/\nu$. From the definition of LLM above, $\kappa \propto B^{-1/2}$ since $E_{Coulomb} \propto 1/l_B \propto \sqrt{B}$ and $E_{cyclotron} \propto B$. Therefore, a sample with low density will have $\nu = 5/2$ occurring at relatively low magnetic fields, which corresponds to high LLM.

7.2   Tuning the Electron–Electron Interactions Through Quantum Well Width

The second parameter taking into account the form of electron–electron interactions in a realistic sample is the adimensional well width, or effective well width. In a realistic model of a sample, the wavefunction in the quantum well is on the order of tens of nanometers. Accounting for finite width softens the form of the Coulomb interaction in the problem, changing the nature of the short range interactions between the particles [8]. Taking the finite width into account can lead to a more accurate calculation of the system ground state, especially at the enigmatic $\nu = 5/2$. It has been found that finite width is in fact necessary for stabilizing the Pfaffian ground state at $\nu = 5/2$ [8–12].

Experimentally, finite width is encoded in an adimensional parameter, w/l_B, where w is the quantum well width. Similarly to LLM, the adimensional width parameter $w/l_B \propto \sqrt{B}$ is a parameter that can be tuned by adjusting the electron density within a single sample, or, alternatively, by exploring different samples grown with different quantum well widths. Indeed, the effect of LLM and finite well width on the $\nu = 5/2$ FQHS has already been studied in detail, both theoretically and in experimentally in low-density samples [3, 4, 6–10, 13–19].

Because different states might exist at different κ and w/l_B, the possibility of a phase transition between different ground states at a fixed filling factor is a very real one. In fact, a phase transition from Pfaffian to Anti-Pfaffian driven by Landau level mixing and different effective well widths is predicted [6, 7, 9, 10]. There is even work on the effect of Landau level mixing on nematicity at $\nu = 5/2$ [20]. Two recent demonstrations of numerically obtained phase diagrams of $\nu = 5/2$ are given in Fig. 7.1, which take into account the effect of Landau level mixing and finite well width. Panels (a) and (b) are from Ref. [6], using a disk geometry, with a disk of size d. Panel (a) assumes zero width, while panel (b) takes into account finite width. The red and green regions are regions where the Pfaffian and Anti-Pfaffian ground states, respectively, are stabilized. It can be seen that finite width broadens these regions. Panel (c) (Ref. [7]) is a calculation using spherical geometry and plots the phases directly in $\kappa - w/l_B$ phase space. The darkest blue region is found to be most likely to support the Pfaffian ground state. These are just two recent examples highlighting the sensitivity of possible ground states to electron–electron interactions.

7.3   The Role of Electron–Electron Interactions in the Fractional Quantum Hall State-to-Nematic Phase Transition

Because in our experiment we changed the magnetic field at which $\nu = 5/2$ and $\nu = 7/2$ arise, we have changed the degree of electron–electron interactions, quantified by the LLM parameter κ and w/l_B, at these states. We therefore propose that the observation of a FQHS-to-nematic quantum critical point at both $\nu = 5/2$ and $\nu = 7/2$ at the same critical magnetic field may be due to the tuning of the electron–electron interaction. We think that this interaction is tuned by the pressure through changing the electron density. The competition of the FQHS and of the nematic hinges on a delicate energy balance of these phases near the quantum critical point. We think that, by tuning the pressure, we access a combination of κ and w/l_B, which in the spirit of Ref. [8], stabilizes the nematic phase. To examine this idea, we plot our measured points in $\kappa - w/l_B$ space, shown in Fig. 7.2. As we tune the pressure, in the $\kappa - w/l_B$ space, we sample the curves shown in this figure. At the critical pressure of the FQHS-to-nematic transition we find $\kappa_c^{5/2} = 1.95$, $w/l_{B,c}^{5/2} = 1.62$ and $\kappa_c^{7/2} = 1.90$, $w/l_{B,c}^{7/2} = 1.63$. Here we took into account the pressure dependence of the effective mass and dielectric constant [22, 23]. Indeed,

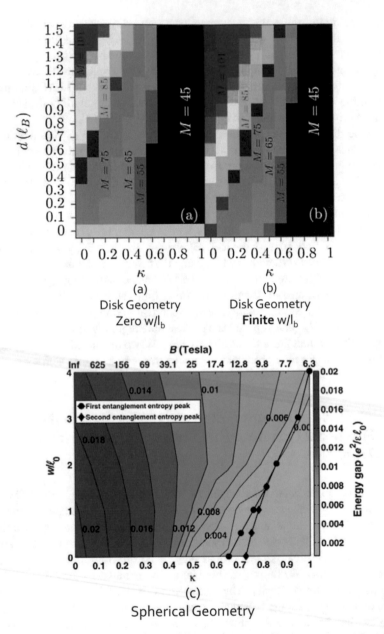

Fig. 7.1 Examples of numerically calculated phase diagrams of states at $v = 5/2$. (**a**) This calculation uses disk geometry, with disk of size d. The plot is in the phase space of d/l_B and Landau level mixing assumes zero width w/l_B. The Pfaffian is found to be stabilized in the red region, while the Anti-Pfaffian is stabilized in green. From Ref. [6]. (**b**) The same plot, but accounting for finite width. This has the effect of broadening the Pfaffian and Anti-Pfaffian regions. From Ref. [6] reprinted figure with permission from A. Tylan-Tyler and Y. Lyanda-Geller, Phys. Rev.B **91**, 205404 (2015). Copyright 2015 by the American Physical Society. (**c**) $\kappa - w/l_B$ phase diagram calculated using a spherical geometry. The dark blue region shows the area of phase space where the Pfaffian ground state is stabilized. From Ref. [7]

Fig. 7.2 The Landau level mixing parameter κ and the adimensional effective well width w/l_B of the FQHSs (open circles) and nematic phases (closed circles) at (**a**) $\nu = 5/2$ and (**b**) $\nu = 7/2$. These are calculated for sample 2 under pressure (blue) and sample 3 (pink star) at ambient pressure, discussed below. The green squares represent the extrapolated critical points of the FQHS-to-nematic transitions in sample 3, and the orange squares represent the extrapolated critical points of the nematic-to-Fermi liquid transition in sample 3. Data from Ref. [21]

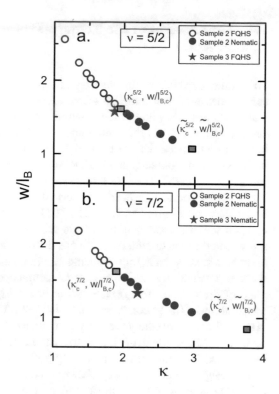

these critical values are nearly the same for both states in the second Landau level, which is indicative that the degree of electron–electron interaction attained here is a universal one for triggering the FQHS-to-nematic transition in the second Landau level.

It is interesting to note that in sample 2 the nematic develops at $\nu = 5/2$ for pressures for which the electron density is in the range of $10.6–6.3 \times 10^{10}\,\mathrm{cm}^{-2}$. Such densities have already been accessed, but the nematic at $\nu = 5/2$ was not observed [1, 15, 18, 19]. Since samples from Refs. [1, 19] had a wider quantum well than our samples, the nematic in them either does not develop or it forms at a yet unknown critical κ and w/l_B parameters. The other two samples, however, had quantum wells of the same width as our samples [15, 18]. In one of these samples the densities necessary for the nematic, lower than $10.6 \times 10^{10}\,\mathrm{cm}^{-2}$, have not been studied [18]. In the other 30 nm quantum well sample the FQHS at $\nu = 5/2$ is seen down to a density $\approx 1.25 \times 10^{10}\,\mathrm{cm}^{-2}$, but the nematic at $\nu = 5/2$ was not seen at $9.5 \times 10^{10}\,\mathrm{cm}^{-2}$ [15]. Possible reasons for the absence of the nematic in Ref. [15] are disorder effects or effects due to the asymmetric shape of the wavefunction in the direction perpendicular to the plane of the electrons in gated samples. Resistance anisotropy at $\nu = 7/2$ was observed in 60 nm quantum well sample having a density of $5 \times 10^{10}\,\mathrm{cm}^{-2}$, providing an important clue on the influence of the width of the quantum well [1]. No data is available at $\nu = 7/2$ in Refs. [15, 18].

7.4 Observation of the Nematic Phase at $v = 7/2$ at Ambient Pressure

To test the relevance of the electron–electron interactions, we investigate a second sample measured at ambient pressure, but in which the electron–electron interaction was tuned near its value at the quantum critical point obtained in the pressurized sample. Sample 3 has the same width of the quantum well as sample 2 (30 nm), but it has a reduced density of $n = 10.9 \times 10^{10}\,\mathrm{cm}^{-2}$. It is a $4 \times 4\,\mathrm{mm}^2$ sample with indium/tin contacts, and we measure it at very low temperatures in our ^{3}He immersion cell [24]. The ^{3}He immersion cell may hold a sample and be filled with liquid ^{3}He, in order to powerfully thermalize the sample. A full description of the immersion cell is given in Ref. [24], but I will summarize its structure here. The cell is made of plastic, and the sample sits inside the cell at the center. Its contacts are soldered to silver sinters which are filled with a fine silver powder that has an enormous surface area, thus reducing Kapitza resistance and massively enhancing thermal conductivity [25]. The cell is mounted on a tail to the mixing chamber plate of the dilution refrigerator. When the refrigerator is cooled to milliKelvin temperatures, ^{3}He is condensed into this cell. The sample is surrounded by the ^{3}He, cooling it extremely well. By immersing the sample in the ^{3}He and having its wires soldered to silver sinters, the sample and the electrons are both very well thermalized. This allows extremely fragile electron ground states to form.

By design, the density was picked in such a way that the parameters κ and w/l_B calculated at $v = 7/2$ fall in the range of the nematic (shown as a pink star in Fig. 7.2). We note that data points for sample 3 in Fig. 7.2 are slightly off the curve for sample 2 since pressure corrections of the mass and dielectric strength are no longer needed. Magnetoresistance traces for this sample are shown in Fig. 7.3. At $v = 7/2$ we indeed observe an extremely large resistance anisotropy. Furthermore, at $v = 5/2$ we observe a FQHS, consistent with the κ and w/l_B parameters being just outside the range for the nematic. We note that the resistance anisotropy of sample 3 at $v = 7/2$ greatly exceeds that in sample 2 shown in Fig. 6.1b, c because of the much lower $T \approx 4.5\,\mathrm{mK}$ electronic temperatures achieved in the ^{3}He immersion cell [24], as compared to $T \approx 12\,\mathrm{mK}$ in the pressure cell.

Taken together, there is compelling evidence that the nematic phase is stabilized in the second orbital Landau level at ambient pressure when the electron–electron interaction is tuned via the parameters κ and w/l_B, to the stability range of the nematic. We emphasize that, according to our findings, the numerical values of the critical parameters of the FQHS-to-nematic transition are valid only for $v = 5/2$ and $7/2$ in the second orbital Landau level and are dependent on parameters such as the width of the quantum well.

Lastly, in Fig. 7.4, we show the quantized Hall resistance at about $T = 12\,\mathrm{mK}$ in the unpressurized sample, sample 3, at $v = 5/2$. This is likewise evidence that $v = 5/2$ in this sample is a well quantized quantum Hall state. Where there is a nematic phase at $v = 7/2$, we observe mixing effects due to the large value of R_{xx}, much like that observed in Fig. 5.2c.

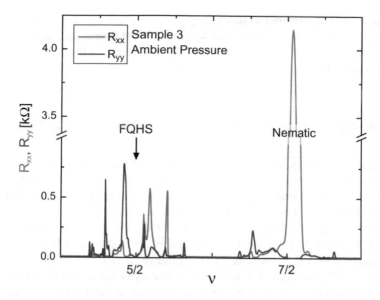

Fig. 7.3 Longitudinal resistance traces in Sample 3, the unpressurized sample, at a base temperature around $T \approx 4.5$ mK. A well-defined FQHS is at $\nu = 5/2$, while the nematic phase appears around $\nu = 7/2$. Data from Ref. [21]

Fig. 7.4 The Hall resistance in the unpressurized sample 3 around $\nu = 5/2$, showing the quantized resistance of this FQHS. The temperature is about $T = 12$ mK

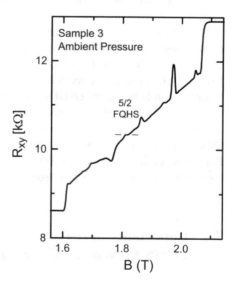

Certain anomalies, possibly due to the effect of fluctuations near the critical point, develop in sample 3. These can be seen in Fig. 7.3. The resistance near $\nu = 5/2$ is not isotropic in the vicinity of $\nu = 5/2$ and data at $\nu \approx 2.42$ suggests a nematic which is not centered at half-filling. Furthermore, resistance anisotropy in the upper spin branch is not exactly centered to $\nu = 7/2$. Since the mean field approach

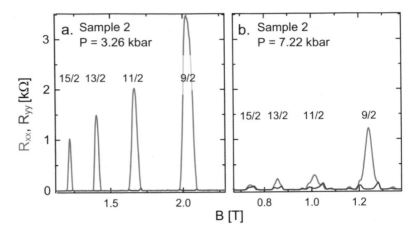

Fig. 7.6 The effect of pressure on the third and fourth Landau levels in the 2DES, in sample 2 at $T = 12$ mK. (**a**) At lowest pressures, here represented at $P = 3.26$ kbar, the anisotropy of the nematic phase is large and it is very robust. (**b**) At higher pressures, here shown at $P = 7.22$ kbar, the nematic phase is highly suppressed in the third and higher Landau levels, and continues to grow less robust with pressure

7.7   Conclusion

We have demonstrated the crucial role of electron–electron interactions, encoded within Landau level mixing and effective quantum well width, in driving a transition from the FQHS to the nematic phase in the second Landau level. The nematic phase at $\nu = 7/2$ in the unpressurized sample 3 arises at a magnetic field well within the range of magnetic fields at which the nematic phase is stabilized in the pressurized samples. These results suggest the existence of a universal critical Landau level mixing parameter and adimensional effective well width for the FQHS-to-nematic transition in the second Landau level in 30 nm quantum well samples. The study of samples of different well widths under pressure may further reveal a universal dependence of the grown quantum well width of the critical Landau level mixing parameter and adimensional effective well width for the stabilization of the nematic phase in the second Landau level.

References

1. W. Pan, A. Serafin, J.S. Xia, L. Yin, N.S. Sullivan, K.W. Baldwin, K.W. West, L.N. Pfeiffer, D.C. Tsui, Phys. Rev. B. **89**, 241302 (2014)
2. D. Yoshioka, J. Phys. Soc. Jpn. **55**, 885 (1986)
3. M.P. Zaletel, R.S.K. Mong, F. Pollmann, E.H. Rezayi, Phys. Rev. B **91**, 045115 (2015)
4. E.H. Rezayi, S.H. Simon, Phys. Rev. Lett. **106**, 116801 (2011)
5. S.Y. Lee, V.W. Scarola, J.K. Jain, Phys. Rev. Lett. **87**, 256803 (2001)

6. A. Tylan-Tyler, Y. Lyanda-Geller, Phys Rev. B. **91**, 205404 (2015)
7. K. Pakrouski, M.R. Peterson, T. Jolicoeur, V.W. Scarola, C. Nayak, M. Troyer, Phys. Rev. X **5**, 021004 (2015)
8. E.H. Rezayi, F.D.M. Haldane, Phys. Rev. Lett. **84**, 4685 (2000)
9. M.R. Peterson, T. Jolicoeur, S. Das Sarma, Phys. Rev. Lett. **101**, 016807 (2008)
10. Z. Papić, N. Regnault, S. Das Sarma, Phys. Rev. B. **80**, 201303 (2009)
11. X. Wan, Z. Hu, E. Rezayi, K. Yang, Phys. Rev. B. **77**, 165316 (2008)
12. H. Wang, D.N. Sheng, F.D.M. Haldane, Phys. Rev. B. **80**, 241311 (2009)
13. A. Wójs, J.J. Quinn, Phys. Rev. B **74**, 235319 (2006)
14. A. Wójs, C. Toke, J.K. Jain, Phys. Rev. Lett. **105**, 096802 (2010)
15. J. Nuebler, V. Umansky, R. Morf, M. Heiblum, K. von Klitzing, J. SMET Phys. Rev. B **81**, 035316 (2010)
16. E.H. Rezayi, Phys. Rev. Lett. **119**, 026801 (2017)
17. J.-S. Jeong, K. Park, Phys. Rev. B **91**, 195119 (2015)
18. J.D. Watson, G.A. Csáthy, M.J. Manfra, Phys. Rev. Appl. **3**, 064004 (2015)
19. N. Samkharadze, D. Ro, L.N. Pfeiffer, K.W. West, G.A. Csáthy, Phys. Rev. B **96**, 085105 (2017)
20. P.M. Smith, M.P. Kennett, J. Phys. Condens. Matter **24**, 055601 (2012)
21. K.A. Schreiber, N. Samkharadze, G.C. Gardner, Y. Lyanda-Geller, M.J. Manfra, L.N. Pfeiffer, K.W. West, G.A. Csáthy, Nat. Commun. **9**, 2400 (2018)
22. D.K. Maude, J.C. Portal, Parallel transport in low-dimensional semiconductor structures. Semicond. Semimetals **55**, 1–43 (1998)
23. Z. Wasilewski, R.A. Stradling, Semicond. Sci. Technol. **1**, 264 (1986)
24. N. Samkharadze, A. Kumar, M.J. Manfra, L.N. Pfeiffer, K.W. West, G.A. Csáthy, Rev. Sci. Inst. **82**, 053902 (2011)
25. F. Pobell, *Matter and Methods at Low Temperatures*, 2nd edn. (Springer, Berlin, 1996)
26. A.A. Koulakov, M.M. Fogler, B.I. Shklovskii, Phys. Rev. Lett. **76**, 499 (1996)
27. R. Moessner, J.T. Chalker, Phys. Rev. B. **54**, 5006 (1996)
28. N. Samkharadze, K.A. Schreiber, G.C. Gardner, M.J. Manfra, E. Fradkin, G.A. Csáthy, Nat. Phys. **12**, 191 (2016)
29. K.A. Schreiber, N. Samkharadze, G.C. Gardner, R.R. Biswas, M.J. Manfra, G.A. Csáthy, Phys. Rev. B **96**, 041107 (2017)
30. Y. You, G.Y. Cho, E. Fradkin, Phys. Rev. B **93**, 205401 (2016)
31. Z. Zhu, I. Sodemann, D.N. Sheng, L. Fu, Phys. Rev. B **95**, 201116 (2017)
32. A. Mesaros, M.J. Lawler, E.-A. Kim, Phys. Rev. B **95**, 125127 (2017)
33. K. Lee, J. Shao, E.A. Kim, F.D.M. Haldane, E.H. Rezayi, Phys. Rev. Lett. **121**, 147601 (2018)
34. J.H. Davies, *The Physics of Low-Dimensional Semiconductors* (Cambridge University Press, Cambridge, 1998)
35. M.M. Fogler, Stripe and bubble phases in quantum hall systems. *High Magnetic Fields Lecture Notes in Physics*, vol. 595, pp. 98–138 (Springer, Berlin, 2002)

Printed in the United States
By Bookmasters